THE FIVE AGES

OF THE UNIVERSE

INSIDE THE PHYSICS OF ETERNITY

FRED ADAMS AND
GREG LAUGHLIN

A Touchstone Book
Published by Simon & Schuster
New York London Toronto Sydney Singapore

TOUCHSTONE
Rockefeller Center
1230 Avenue of the Americas
New York, NY 10020

First Touchstone Edition 2000

TOUCHSTONE and colophon are registered trademarks
of Simon & Schuster, Inc.

Designed by Carla Bolte

Manufactured in the United States of America

10 9 8 7 6 5 4 3 2

The Library of Congress has cataloged the Free Press edition as follows:

Adams, Fred, date
The five ages of the universe : inside the physics of eternity /
Fred Adams, Greg Laughlin.
p. cm.
Includes bibliographical references and index.
1. Cosmology. I. Laughlin, Greg. II. Title.
QB981.A36 1999
523.1—dc21 99-18139
CIP

ISBN 0-684-85422-8 (alk. paper)
0-684-86576-9 (Pbk)

CONTENTS

PREFACE

The long-term fate of our universe has always drawn our fascination. As astrophysicists, we are used to thinking about both the origin and fate of astronomical objects, such as stars, galaxies, and often the universe as a whole. During the twentieth century, and especially within the past two decades, physics, astronomy, and cosmology have advanced rapidly and we can now understand our universe with unprecedented clarity.

Most cosmological work has focused on the past history of the universe, and rightly so. The big bang theory and its modifications provide a highly successful paradigm for describing the origin of the universe, and recent astronomical data place this endeavor on a solid scientific foundation. Parallel developments in stellar evolution, star formation, and galaxy formation have rounded out our detailed understanding of the present-day universe. In recent years, the inventory of our universe has been bolstered further by the astronomical discovery of black holes, brown dwarfs, and planets orbiting other stars.

Against this background, late in the year 1995, we began a detailed scientific inquiry into the universe of the future, including the stars, galaxies, and other astrophysical entities living within it. Our research was motivated by a convergence of events. We felt that the time was right to reopen this issue. In spite of the recent progress concerning the history of the universe, relatively little attention had been given to the future universe. Landmark papers by Martin Rees, Jamal Islam, and Freeman Dyson were published in the 1960s and

1970s, and a small flurry of work outlining the consequences of proton decay followed in the early 1980s. With the greater perspective provided by recent progress in both physics and astronomy, however, we could take a much more detailed glimpse into the future. Past studies dealt mostly with large-scale cosmological issues, with little focus on stars and stellar objects. In fact, the long-term evolution of the lowest-mass stars, which live far longer than the current age of the universe, had not yet been calculated. This project was facilitated by Laughlin's arrival at the University of Michigan to work with Adams, thus beginning our joint effort to peer into the darkness. As an added incentive, the University of Michigan held a "theme semester" which focused on "Death, Extinction, and the Future of Humanity." For this occasion, Adams developed a new class on the long-term future of the universe. The relative paucity of reference material provided additional impetus to study the far future.

The results of our initial research work were published as a review article in *Reviews of Modern Physics,* in 1997, with the encouragement and guidance of Sir Martin Rees, the journal editor and the Astronomer Royal of England. Following the completion of the scientific manuscript, the popular media showed enthusiastic interest in the subject and the demand for public lectures was surprisingly high. This unforeseen public curiosity led us to consider writing a popular treatment of this topic, and this book is the result.

Our book differs from many within its class in that it introduces a substantial amount of new scientific material. During the production of this manuscript, we often faced scientific issues that had not been previously studied. For example, what happens if a small red dwarf star enters our solar system and disrupts the orbits of the planets? Can Earth be captured by such an encounter? What are the chances of this event taking place before our Sun dies? To answer these and other questions of this nature, we simply performed the necessary computations as the need arose. As a result, our book contains the results from new, previously unpublished, calculations and other scientific arguments. It thus presents the most complete and detailed treatment of the future to date.

Nevertheless, this book is intended for a general audience, and no previous knowledge of physics or astronomy is required to appreciate this biography of our universe. During this discussion, however, we often need to consider some very large numbers. In most instances, we present both the names for large numbers, such as billions and trillions, as well as the number in scientific notation. For example, a billion can be written as 10^9, a 1 followed by nine zeros; a trillion can be written as 10^{12}, a 1 followed by twelve zeros, and so on. We also introduce a convenient new unit of time called a "cosmological decade." When an interval of time, in years, is expressed in scientific notation, say 10^η years, then the exponent η is the cosmological decade. The universe is now ten billion or 10^{10} years old, for example, which means that we are now living in the 10th cosmological decade. When the universe is ten times older, 10^{11} years old, the universe will experience its 11th cosmological decade. In this book, we describe events of astronomical importance spanning time intervals up to 150 cosmological decades, or 10^{150} years.

Any description of the future universe necessarily contains speculative elements. In this book, we begin with our current knowledge of the laws of physics and our current understanding of astrophysics, and extrapolate forward to construct a vision of the future universe. Our extrapolation continues to lose focus, however, as we progress ever deeper into the dimness of future time. Along the way, we must incorporate the effects of physical and astronomical processes that remain under study. In particular, most of this book outlines future cosmology for a universe that expands forever, which is indeed suggested by current astronomical data. For completeness, however, a short discussion of a recollapsing universe is also included. In addition, we explicitly assume that both proton decay and the Hawking evaporation of black holes will eventually take place. Although both of these long-term processes are predicted in general terms by theoretical physics, they have not yet been experimentally verified.

Almost everything we discuss in this book rests on an additional article of faith. We assume that the laws of physics will continue to hold and will not change with time, at least not until after the time line of this chronicle has run out. Although we have no absolute guarantee that this assumption holds, we also have no compelling reason to doubt it. Many different lines of evidence show that the character of physical law has remained remarkably constant thus far in the history of the universe, and we have no indications that this trend will not continue. For example, studies of the early universe in the context of the big bang theory strongly suggest that the physical laws which describe nature are both fairly well understood and have held constant from very early times to the present. When we assume that the laws of physics will not change, we are referring to the actual laws themselves, and not our understanding of them. Improvements in the understanding of physics will almost certainly lead to alterations, both large and small, in the vision of the future presented here.

This book tells the story of our universe, from its singular inception at the big bang to its disintegrated slide into the far future. This saga is punctuated by stellar explosions, collisions, and a multitude of other astrophysical disasters. At first glance, the unending dilapidation of the universe may seem like a dismal or depressing prospect. To our minds, however, the ever changing characteristics of the universe provide us with a much grander perspective, a wider view of the cosmos and our place within it. Our hope is that readers will glean a better understanding of the history of our universe, what it contains, how it works, and where it might be headed into the future. From this enriched frame of reference, one can attain a much greater appreciation of our "ordinary" interactions within the comfortable universe of the present epoch.

During the writing of this book, we have benefited greatly from the assistance of many people. Our agent, Lisa Adams, was instrumental in getting this project off the ground. Our editor, Stephen Morrow from The Free Press, skillfully guided the book through the many obstacles necessary for publication. Along the way, many interested friends and colleagues provided much appreciated help and criticism, including Peter Bodenheimer, Noel Brewer, Myron Campbell, Gus Evrard, Alex Filippenko, Margaret Gibbons, Gordy Kane, Mark Laughlin, Ron Mallett, Manasse Mbonye, Sally Pobojewski, Yopie Prins, Roy Rappaport, Martin Rees, Michal Rozyczka, Nathan Schwadron, J. Allyn Smith, David Spergel, and Lisa Stillwell. Finally, we would like to thank Laurel Taylor for preparing an elegant and informative set of diagrams.

Fred Adams and Greg Laughlin

October 1998, Ann Arbor, Michigan

INTRODUCTION

A GUIDE TO THE BIG PICTURE, FUNDAMENTAL PHYSICAL LAW, WINDOWS OF SPACE AND TIME, THE GREAT WAR, AND EXTREMELY BIG NUMBERS.

January 1, 7,000,000,000 A.D., Ann Arbor:

The New Year rings in little cause for celebration. Nobody is present even to mark its passing. Earth's surface is a torrid unrecognizable wasteland. The Sun has swelled to enormous size, so large that its seething red disk nearly fills the daytime sky. The planet Mercury and then Venus have already been obliterated, and now the tenuous outer reaches of the solar atmosphere are threatening to overtake the receding orbit of Earth.

Earth's life-producing oceans have long since evaporated, first into a crushing, sterilizing blanket of water vapor, and then into space entirely. Only a barren rocky surface is left behind. One can still trace the faint remains of ancient shorelines, ocean basins, and the low eroded remnants of the continents. By noon, the temperature reaches nearly 3000 degrees Fahrenheit, and the rocky surface begins to melt. Already, the equator is partly ringed by a broad glowing patchwork of lava, which cools to form a thin gray crust as the distended Sun eases beneath the horizon each night.

A patch of the surface which once cradled the forested moraines of southeastern Michigan has seen a great deal of change over the intervening billions of years. What was once the North American continent has long since been torn apart by the geologic rift which opened from Ontario to Louisiana and separated the old stable continental platform to produce a new expanse of sea floor. The sedimented, glaciated remains of Ann Arbor were covered by lava which arrived from nearby rift volcanos by coursing through old river channels. Later, the hardened lava and the underlying sedimentary rock were thrust up into a mountain chain as a raft of islands the size of New Zealand collided with the nearby shoreline.

Now, the face of an ancient cliff is weakened by the Sun's intense heat. A slab of rock cleaves off, causing a landslide and exposing a perfectly preserved fossil of an oak leaf. This trace of a distant verdant world slowly melts away in the unyielding heat. Soon the entire Earth will be glowing a sullen, molten red.

This scenario of destruction is not the lurid opening sequence from a grade B movie, but rather a more or less realistic description of the fate of our planet as the Sun ends its life as a conventional star and expands to become a red giant. The catastrophic melting of Earth's surface is just one out of a myriad of events that are waiting to occur as the universe and its contents grow older.

Right now, the universe is still in its adolescence with an age of ten to fifteen billion years. As a result, not enough time has elapsed for many of the more interesting astronomical possibilities to have played themselves out. As the distant future unfolds, however, the universe will gradually change its character and will support an ever changing variety of astonishing astrophysical processes. This book tells the biography of the universe, from beginning to end. It is the story of the familiar stars of the night sky slowly giving way to bizarre frozen stars, evaporating black holes, and atoms the size of galaxies. It is a scientific glimpse at the face of eternity.

FOUR WINDOWS TO THE UNIVERSE

Our biography of the universe, and the study of astrophysics in general, plays out on four important size scales: planets, stars, galaxies, and the universe as a whole. Each of these scales provides a different type of window to view the properties and evolution of nature. On each of these size scales, astrophysical objects go through life cycles, beginning with a formation event analogous to birth and often ending with a specific and deathlike closure. The end can come swiftly and violently; for example, a massive star ends its life in a spectacular supernova explosion. Alternatively, death can come tortuously slowly, as with the dim red dwarf stars, which draw out their lives by slowly fading away as white dwarfs, the cooling embers of once robust and active stars.

On the largest size scale, we can view the universe as a single evolving entity and study its life cycle. Within this province of *cosmology,* a great deal of scientific progress has been accomplished in the past few decades. The universe has been expanding since its conception during a violent explosion—the big bang itself. The big bang theory describes the subsequent evolution of the universe over the last ten to fifteen billion years and has been stunningly successful in explaining the nature of our universe as it expands and cools.

The key question is whether the universe will continue to expand forever or halt its expansion and recollapse at some future time. Current astronomical data strongly suggest that the fate of our universe lies in continued expansion, and most of our story follows this scenario. However, we also briefly lay out the consequences of the other possibility, the case of the universe recollapsing in a violent and fiery death.

Beneath the grand sweep of cosmology, at a finer grain of detail, are the galaxies, such as our Milky Way. These galaxies are large and somewhat loosely knit collections of stars, gas, and other types of matter. Galaxies are not distributed randomly throughout the universe, but rather they are woven into a tapestry by gravity. Some aggregates of galaxies have enough mass to be bound together by gravitational forces, and these galaxy clusters can be considered as independent astrophysical objects in their own right. In addition to belonging to clusters, galaxies are loosely organized into even larger structures that resemble filaments, sheets, and walls. The patterns formed by galaxies on this size scale are collectively known as the *large-scale structure of the universe.*

Galaxies contain a large fraction of the ordinary mass in the universe and are well separated from each other, even within their clusters. This separation is so large that galaxies were once known as "island universes." Galaxies also play the extremely important role of marking the positions of space-time. As the universe expands, the galaxies act as beacons in the void that allow us to observe the expansion.

It is difficult to comprehend the vast emptiness of our universe. A typical galaxy fills only about one-millionth of the volume of space that contains the galaxy, and the galaxies themselves are extremely tenuous. If you were to fly a spaceship to a random point in the universe, the chances of landing within a galaxy are about one in a million at the present time. These odds are already not very good, and in the future they will only get worse, because the universe is expanding but the galaxies are not. Decoupled from the overall expansion of the universe, the galaxies exist in relative isolation. They are the homes of most stars in the universe, and hence most planets. As a result, many of the interesting physical processes in the universe, from stellar evolution to the evolution of life, take place within galaxies.

Just as they do not thickly populate space, the galaxies themselves are mostly empty. Very little of the galactic volume is actually filled by the stars, although galaxies contain billions of them. If you were to drive a spaceship to a random point in our galaxy, the chances of landing within a star are extremely small, about one part in one billion trillion (one part in 10^{22}). This emptiness of galaxies tells us much about how they have evolved and how they will endure in the future. Direct collisions between the stars in a galaxy are exceedingly rare. Consequently, it takes a very long time, much longer than the current age of the universe, for stellar collisions and other encounters to affect the structure of a galaxy. As we shall see, these collisions become increasingly important as the universe grows older.

The space between the stars is not entirely empty. Our Milky Way is permeated with gas of varying densities and temperatures. The average density is only about one particle (one proton) per cubic centimeter and the temperature ranges from a cool 10 degrees kelvin to a seething million degrees. At low temperatures, about 1 percent of the material lives in solid form, in tiny rocky dust particles. This gas and dust that fill in the space between stars are collectively known as the *interstellar medium.*

The stars themselves give us the next smaller size scale of importance. Ordinary stars, objects like our Sun which support themselves through nuclear fusion in their interiors, are now the cornerstone of astrophysics. The stars make up the galaxies and generate most of the visible light seen in the universe. Moreover, stars have shaped the current inventory of the universe. Massive stars have forged almost all of the heavier elements that spice up the cosmos, including the carbon and oxygen required for life. Most of what makes up the ordinary matter of everyday experience—books, cars, groceries—originally came from the stars.

But these nuclear power plants cannot last forever. The fusion reactions that generate energy in stellar interiors must eventually come to an end as the nuclear fuel is exhausted. Stars with masses much larger than our Sun burn themselves out within a relatively brief span of a few million years, a lifetime one thousand times shorter than the present age of the universe. At the other end of the range, stars with masses much less than that of our Sun can last for trillions of years, about one thousand times the current age of the universe.

When stars end the nuclear burning portion of their lives, they do not disappear altogether. In their wake, stars leave behind exotic condensed objects collectively known as *stellar remnants.* This cast of degenerate characters includes brown dwarfs, white dwarfs, neutron stars, and black holes. As we shall see, these strange leftover remnants will exert an increasingly important and eventually dominant role as the universe ages and the ordinary stars depart from the scene.

The planets provide our fourth and smallest size scale of interest. They come in at least two varieties: relatively small rocky bodies like our Earth, and larger gaseous giants like Jupiter and Saturn. The last few years have seen an extraordinary revolution in our understanding of planets. For the first time in history, planets in orbit about other stars have been unambiguously detected. We now know with certainty that planets are relatively commonplace in the galaxy, and not just the outcome of some rare or special event which occurred in our solar system. Planets do not play a major role in the evolution and dynamics of the universe as a whole. They are important because they provide the most likely environments for life to evolve. The long-term fate of planets thus dictates the long-term fate of life—at least the life-forms with which we are familiar.

In addition to planets, solar systems contain many smaller objects, such as asteroids, comets, and a wide variety of moons. As in the case of planets, these bodies do not play a major role in shaping the evolution of the universe as a whole, but they do have an important impact on the evolution of life. The moons orbiting the planets provide another possible environment for life to thrive. Comets and asteroids are known to collide with planets on a regular basis. These impacts, which can drive global climatic changes and extinctions of living species, are believed to have played an important role in shaping the history of life here on Earth.

THE FOUR FORCES OF NATURE

Nature can be described by four fundamental forces which ultimately drive the dynamics of the entire universe: gravity, the electromagnetic force, the strong nuclear force, and the weak nuclear force. All four of these forces play significant roles in the biography of the cosmos. They have helped shape our present-day universe and will continue to run the universe throughout its future history.

The first of these forces, gravity, is the closest to our everyday experience and is actually the weakest of the four. Since it has a long range and is always attractive, however, gravity dominates the other forces on sufficiently large size scales. Gravity holds objects to Earth, and holds Earth in its orbit around the Sun. Gravity keeps the stars intact and drives their energy generation and evolution. Ultimately, it is the force responsible for forming most structures in the universe, including galaxies, stars, and planets.

The second force is the electromagnetic force, which includes both electric and magnetic forces. At first glance, these two forces might seem different, but at the fundamental level they are revealed to be aspects of a single underlying force. Although the electromagnetic force is intrinsically much stronger than gravity, it has a much smaller effect on large size scales. Positive and negative charges are the source of the electromagnetic force and the universe appears to have an equal amount of each type of charge. Because the forces created by charges of opposite sign have opposite effects, the electromagnetic force tends to cancel itself out on large size scales that contain many charges. On small size scales, in particular within atoms, the electromagnetic force plays a vitally important role. It is ultimately responsible for most of atomic and molecular structure, and hence is the driving force in chemical reactions. At the fundamental level, life is governed by chemistry and the electromagnetic force.

The electromagnetic force is a whopping 10^{40} times stronger than the gravitational force. One way to grasp this overwhelming weakness of gravity is to imagine an alternate universe containing no charges and hence no electromagnetic forces. In such a universe, ordinary atoms would have extraordinary properties. With only gravity to bind an electron to a proton, a hydrogen atom would be larger than the entire observable portion of our universe.

The strong nuclear force, our third fundamental force of nature, is responsible for holding atomic nuclei together. The protons and the neutrons are held together in the nucleus by this force. Without the strong force, atomic nuclei would explode in response to the repulsive electric forces between the positively charged protons. Although it is intrinsically the strongest of the four forces, the strong force has a very short range of influence. Not by coincidence, the range of the strong nuclear force is about the size of a large atomic nucleus, about ten thousand times smaller than the size of an atom (about 10 fermi or 10^{-12} centimeters). The strong force drives the process of nuclear fusion, which in turn provides most of the energy in stars and hence in the universe at the present epoch. The large magnitude of the strong force in comparison with the electromagnetic force is ultimately the reason why nuclear reactions are much more powerful than chemical reactions, by a factor of a million on a particle-by-particle basis.

The fourth force, the weak nuclear force, is perhaps the farthest removed from the public consciousness. This rather mysterious weak force mediates the decay of neutrons into protons and electrons, and also plays a role in nuclear fusion, radioactivity, and the production of the elements in stars. The weak force has an even shorter range than the strong force. In spite of its weak strength and short range, the weak force plays a surprisingly important role in astrophysics. A substantial fraction of the total mass of the universe is most likely made up of weakly interacting particles, in other words, particles that interact only through the weak force and gravity. Because such particles tend to interact on very long time scales, they play an increasingly important role as the universe slowly cranks through its future history.

THE GREAT WAR

A recurring theme throughout the life of the universe is the continual struggle between the force of gravity and the tendency for physical systems to evolve toward more disorganized conditions. The amount of disorder in a physical system is measured by its *entropy* content. In the broadest sense, gravity tends to pull things together and thereby organizes physical structures. Entropy production works in the opposite direction and acts to make physical systems more disorganized and spread out. The interplay between these two competing tendencies provides much of the drama in astrophysics.

Our Sun provides an immediate example of this ongoing struggle. The Sun lives in a state of delicate balance between the action of gravity and entropy. The force of gravity holds the Sun together and pulls all of the solar material toward the center. In the absence of competing forces, gravity would rapidly crush the Sun into a black hole only several kilometers across. This disastrous collapse is prevented by pressure forces which push outward to balance the gravitational forces and thereby support the Sun. The pressure that holds up the Sun ultimately arises from the energy of nuclear reactions taking place in the solar interior. These reactions generate both energy and entropy, leading to random motions of the particles in the solar core, and ultimately supporting the structure of the entire Sun.

On the other hand, if the force of gravity was somehow shut off, the Sun would no longer be confined and would quickly expand. This dispersal would continue until the solar material was spread thinly enough to match the very low densities of interstellar space. The rarefied ghost of the Sun would then be several light-years across, about 100 million times its present size.

The evenly matched competition between gravity and entropy allows the Sun to exist in its present state. If this balance is disrupted, and either gravity or entropy overwhelms the other, then the Sun could end up either as a small black hole or a very diffuse wisp of gas. This same state of affairs—a balance of gravity and entropy—determines the structure of all the stars in the sky. The fierce rivalry between these two opposing tendencies drives stellar evolution.

This same general theme of competition underlies the formation of astronomical structures of every variety, including planets, stars, galaxies, and the large-scale structure of the universe. The existence of these astrophysical systems is ultimately due to gravity, which acts to pull material together. Yet in each case, the tendency toward gravitational collapse is opposed by disruptive forces. On every scale, the relentless competition between gravity and entropy ensures that a victory is often temporary, and never entirely complete. For example, the formation of astrophysical structures is never completely efficient. Successful formation events mark local triumphs for gravity, whereas failed incidences of formation represent victories for disorganization and entropy.

This great war between gravity and entropy determines the long-term fate and evolution of astrophysical objects such as stars and galaxies. After a star has burned through all of its nuclear fuel, for example, it must adjust its internal structure accordingly. Gravity pulls the star inwards, whereas the tendency for increasing entropy favors dispersal of the stellar material. The subsequent battle can have many different outcomes, depending on the mass of the star and its other properties (for example, the rate at which the star spins). As we shall see, this drama will be repeated over and over again, as long as stellar objects populate the universe.

The evolution of the universe itself provides an intensely dramatic example of the ongoing struggle between the force of gravity and entropy. The universe is expanding and becoming more spread out with time. Resisting this evolutionary trend is the force of gravity, which tries to pull the expanding material of the universe back together. If gravity wins this battle, the universe must eventually halt its expansion and begin to recollapse some time in the future. On the other hand, if gravity loses the battle, the universe will continue to expand forever. Which one of these fates lies in our future path depends on the total amount of mass and energy contained within the universe.

THE LIMITS OF PHYSICS

The laws of physics describe how the universe works on a wide range of size scales, from the enormously large to the tremendously small. A high-water mark of human accomplishment is our ability to explain and predict how nature behaves in regimes that are vastly disconnected from our everyday Earthbound experience. Most of this expansion of our horizons has occurred within the past century. Our realm of knowledge has been extended from the largest scale structures of the universe all the way down to subatomic particles. Although this domain of understanding may seem large, we must keep in mind that discussions of physical law cannot be extended arbitrarily far in either direction. The very largest and the very smallest size scales remain beyond the reach of our current scientific understanding.

Our physical picture of the largest size scales in the universe is limited by causality. Beyond a certain maximum distance, information has simply not had time to reach us during the relatively short life of the universe. Einstein's theory of relativity implies that no signals that contain information can travel faster than the speed of light. So, given that the universe has lived for only about ten billion years, no information-bearing signals have had time to travel farther than ten billion light-years. This distance provides a boundary to the part of the universe that we can probe with any kind of physics; this causality boundary is often referred to as the *horizon scale.* Because of the existence of this causality barrier, very little can be ascertained about the universe at distances greater than the horizon scale. This horizon scale depends on the cosmological time. In the past, when the universe was much younger, this horizon scale was correspondingly smaller. As the universe ages, the horizon scale continues to grow.

The cosmological horizon is an extremely important concept that limits the playing field of science. Just as a football game must take place within well defined boundaries, physical processes in the universe are constrained to occur within the horizon at any given time. In fact, the existence of a causal horizon leads to some ambiguity regarding what the term "universe" actually means. The term sometimes refers to only the material that is within the horizon at a given time. In the future, however, the horizon will grow and hence will eventually encompass material that is currently outside our horizon. Is this "new" material part of the universe at the current time? The answer can be yes or no, depending on how you define "the universe." Similarly, there can be other regions of space-time that will never lie within our horizon. For the sake of definiteness, we consider such regions of space-time to belong to "other universes."

On the smallest size scales, the predictive power of physics is also limited, but for an entirely different reason. On size scales smaller than about 10^{-33} centimeters (this scale is known as the *Planck length*), the nature of space-time is very different than on larger length scales. At these tiny size scales, our conventional concepts of space and time no longer apply because of quantum mechanical fluctuations. Physics at this scale must simultaneously incorporate both the *quantum theory* and *general relativity* to describe space and time. The quantum theory implies that nature has a wavelike character at sufficiently small size scales. For example, in ordinary matter, the electrons orbiting the nucleus of an atom display many wave properties. The quantum theory accounts for this waviness. The theory of general relativity holds that the geometry of space itself (along with time: space and time are intimately coupled at this fundamental level) changes in the presence of large amounts of matter, which produce strong gravitational fields. Unfortunately, however, we do not yet have a complete theory that combines both quantum mechanics and general relativity. The absence of such a theory of *quantum gravity* greatly limits what we can say about size scales smaller than the Planck length. As we shall see, this limitation of physics greatly inhibits our understanding of the very earliest times in the history of the universe.

COSMOLOGICAL DECADES

In this biography of the universe, the ten billion years already gone by represent an utterly insignificant fragment of time. We must take up the formidable challenge of establishing a time line which depicts the universally interesting events that are likely to transpire over the next 10^{100} years.

The number 10^{100} is big. Very big. Written down without the benefit of scientific notation, this number consists of a 1 followed by one hundred zeros and it looks like this:

10,000,000,000,000,000,000,000,000,000,000,000,000,000,000,000,
000,000,000,000,000,000,000,000,000,000,000,000,000,000,
000,000,000,000,000.

Not only is the number 10^{100} rather cumbersome to write out, but it is difficult to obtain an accurate feeling for just how tremendously gigantic it is. Attempts to visualize 10^{100} by imagining collections of familiar objects are soon thwarted. For example, the number of grains of sand on all of the beaches in the world is often trotted out as an example of an incomprehensibly large number. However, a rough estimate shows that the total number of sand grains is about 10^{23}, a 1 followed by 23 zeros, a big number but still hopelessly inadequate to the task. How about the number of stars in the sky? The number of stars in our galaxy is close to one hundred billion, again a relatively small number. The number of stars in all the galaxies in our observable universe is about 10^{22}, still far too small. In fact, in the entire visible universe, the total number of protons, the fundamental building blocks of ordinary matter, is only 10^{78}, still a factor of ten billion trillion times too small! The number of years between here and eternity is truly immense.

In order to describe the time scales involved in the future evolution of the universe, without becoming completely bewildered, let's use a new unit of time called a *cosmological decade.* If τ is the time in years, then τ can be written in scientific notation in the form

$$\tau = 10^{\eta} \text{ years,}$$

where η is some number. According to our definition, the exponent η is the number of cosmological decades. For example, the universe is currently only about ten billion years old, which corresponds to 10^{10} years or $\eta = 10$ cosmological decades. In the future, when the universe is 100 billion years old, the time will be 10^{11} years or $\eta = 11$ cosmological decades. The power of this scheme is that each successive cosmological decade represents a tenfold increase in the total age of the universe. The concept of a cosmological decade thus provides us with a way to think about immensely long time spans. Our aggressively large example, the number 10^{100}, thus corresponds to the rather more tractable hundredth cosmological decade, or $\eta = 100$.

We can also use cosmological decades to refer to the very short but very eventful slivers of time which came immediately after the big bang. We simply allow the cosmological decade to be a negative number. With this extension, one year after the big bang corresponds to 10^0 years, or the zeroth cosmological decade. One-tenth or 10^{-1} years is thus cosmological decade -1, and one-hundredth or 10^{-2} years is cosmological decade -2, and so on. The beginning of time corresponds to $\tau = 0$ when the big bang itself took place; in terms of cosmological decades, the big bang is understood to have occurred at the cosmological decade corresponding to negative infinity.

FIVE GREAT ERAS OF TIME

We can organize our current understanding of the past and future history of the universe by defining the following distinct eras of time. As the universe passes from one era to the next, the inventory and character of the universe change rather dramatically, and in some ways almost completely. These eras, which are analogous to geological eras, provide a broad outline of the life of the universe. As time unfolds, a series of natural astronomical disasters shape the universe and drive its subsequent evolution. A newsreel of this story might run like this:

The Primordial Era. $-50 < \eta < 5$. This era encompasses the early phase in the history of the universe. While the universe is less than 10,000 years old, most of the energy density of the universe resides in the form of radiation and this early time is often called the *radiation dominated era.* No astrophysical objects, like stars or galaxies, have been able to form.

During this brief early epoch, many important events took place which served to set the future course of the universe. The synthesis of light elements, such as helium and lithium, occurred during the first few minutes of this Primordial Era. Even closer to the beginning, complex physical processes set up a small excess of ordinary baryonic matter over antimatter. The antimatter annihilated almost completely with most of the matter and left behind the small residue of matter which makes up the universe of today.

In turning back the clock even further, our understanding becomes shakier. At extraordinarily early times, when the universe was unbelievably hot, it seems that very high energy quantum fields drove a period of fantastically rapid expansion and produced very small density fluctuations in the otherwise featureless universe. These tiny fluctuations have survived and grown into galaxies, clusters, and the large-scale structures that populate the universe today.

Near the end of the Primordial Era, the energy density of the radiation became less than the energy density associated with the matter. This crossover took place when the universe was about ten thousand years old. Shortly thereafter, another watershed event took place as the temperature of the universe became cool enough for atoms (or more specifically, hydrogen atoms) to exist. The first appearance of neutral hydrogen atoms is known as *recombination.* After recombination took place, fluctuations in the density of matter in the universe allowed it to grow into clumps without being affected by the pervasive sea of radiation. Familiar astrophysical objects, like galaxies and stars, began forming for the first time.

The Stelliferous Era. $6 < \eta < 14$. Stelliferous means "filled with stars." During this era, most of the energy generated in the universe arises from nuclear fusion in conventional stars. We now live in the middle of the Stelliferous Era, a time period when stars are actively forming, living, and dying.

In the earliest part of the Stelliferous Era, when the universe was only a few million years old, the first generation of stars was born. During the first billion years, the first galaxies appeared and began organizing themselves into clusters and superclusters.

Many freshly formed galaxies experience violent phases in connection with their rapacious central black holes. As the black holes rip apart stars and surround themselves with whirlpool-like disks of hot gas, vast quantities of energy are released. Over time, these *quasars* and *active galactic nuclei* slowly die down.

In the future, near the end of the Stelliferous Era, a key role will be played by the universe's most ordinary stars—the low-mass stars known as red dwarfs. Red dwarf stars have less than half the mass of the Sun, but they are so numerous that their combined mass easily exceeds the mass of all the larger stars in the universe. These red dwarfs are true misers when it comes to fusing hydrogen into helium. They hoard their energy and will still be around ten trillion years from now, after the larger stars have long since exhausted their nuclear fuel and evolved into white dwarfs or exploded as supernovae. The Stelliferous Era comes to a close when the galaxies run out of hydrogen gas, star formation ceases, and the longest-lived (lowest mass) red dwarfs slowly fade away. When the stars finally stop shining, the universe will be about one hundred trillion years old (cosmological decade $\eta = 14$).

The Degenerate Era. $15 < \eta < 39$. After star formation and conventional stellar evolution have ended, most of the ordinary mass in the universe is locked up in degenerate stellar remnants which remain after stellar evolution has run its course. In this context, degeneracy connotes a peculiar quantum mechanical state of matter, rather than a state of moral turpitude. The inventory of degenerates includes brown dwarfs, white dwarfs, neutron stars, and black holes. During the Degenerate Era, the universe looks very different from the way it appears now. No visible radiation from ordinary stars can light up the night skies, warm the planets, or endow galaxies with the faint glow they have today. The universe is colder, darker, and more diffuse.

Nevertheless, events of astronomical interest continually sparkle against the darkness. Chance close encounters scatter the orbits of dead stars, and the galaxies gradually readjust their structure. Some stellar remnants are ejected far beyond the edge of the galaxy, while others fall in towards the center. A rare beacon of light can emerge when two brown dwarfs collide to create a new low-mass star, which will subsequently live for trillions of years. On average, at any given time, a few such stars will be shining in a galaxy the size of our Milky Way. Every so often, as two white dwarfs collide, the galaxy is rocked by a supernova explosion.

During the Degenerate Era, the white dwarfs, which are the most common stellar remnants, contain most of the ordinary baryonic matter in the universe. These white dwarfs sweep up dark matter particles, which orbit the galaxy in an enormous diffuse halo. Once trapped in the interior of a white dwarf, these particles subsequently annihilate and thereby provide an important energy source for the universe. Indeed, the annihilation of dark matter replaces conventional nuclear burning in stars as the dominant energy mechanism. By the 30th cosmological decade ($\eta = 30$) or sooner, however, the supply of the dark matter particles becomes depleted and this avenue of energy generation comes to an end. The matter inventory of the universe is then limited to white dwarfs, brown dwarfs, neutron stars, and dead, widely scattered planets.

At the end of the Degenerate Era, the mass-energy stored within the white dwarfs and neutron stars dissipates into radiation as their constituent protons and neutrons decay. A white dwarf fueled by proton decay generates approximately 400 watts, enough power to run a few light bulbs. An entire galaxy of these erstwhile stars has a total luminosity smaller than one ordinary hydrogen-burning star like the Sun. As the proton decay process grinds to completion, the Degenerate Era draws to a close. The universe—ever darker, ever more rarefied—changes its character yet again.

The Black Hole Era. $40 < \eta < 100$. After the epoch of proton decay, the only stellarlike objects remaining are the black holes. These fantastic objects have such strong gravitational fields that even light cannot escape from their surfaces. The black holes are unaffected by proton decay and survive unscathed through the end of the Degenerate Era.

As the white dwarfs evaporate and disappear, the black holes slowly sweep up material and grow larger. Yet even black holes cannot last forever. They must eventually evaporate away through a very slow quantum mechanical process known as *Hawking radiation.* In spite of their name, black holes are not completely black. In reality, they shine ever so faintly by emitting a thermal spectrum of light and other decay products. After the protons are gone, the evaporation of black holes, almost by default, provides the universe with its primary source of energy. A black hole with the mass of the Sun lasts for about 65 cosmological decades; a large black hole with the mass of a galaxy takes about 98 to 100 cosmological decades to evaporate. All black holes are thus slated for destruction. The Black Hole Era is over when the largest black holes have evaporated.

The Dark Era. $\eta > 101$. After a hundred cosmological decades, the protons have long since decayed and black holes have evaporated. Only leftover waste products from these processes remain: photons of colossal wavelength, neutrinos, electrons, and positrons. An odd parallel exists between the Dark Era and the Primordial Era, when the universe was less than a million years old. In each of these eras, distantly separated in time, no stellarlike objects of any kind are present to generate energy.

In this cold and distant future, activity in the universe has tailed off dramatically. Energy levels are low and the expanses of time are mind-boggling. Electrons and positrons drifting through space encounter one another and occasionally form positronium atoms. These late-forming structures are unstable, however, and their constituent particles must eventually annihilate. Other low-level annihilation events can also take place, albeit very slowly.

In comparison with its profligate past, the universe now lives a relatively conservative and low-profile existence. Or does it? The seeming poverty of this distant epoch could be due to our uncertain extrapolation, rather than an actual slide into senescence.

SURVIVAL OF LIFE

Our society is steeped in the uneasy awareness that human extinction is not a particularly farfetched possibility. Nuclear armageddon, ecological catastrophes, and rampant viruses are among the doomsday prospects thrust forward by the prudent, the paranoid, and the profit minded. But what if we adopt a somewhat outdated, yet decidedly more romantic, outlook of rocket ships, space colonies, and galactic conquest? In such a future, humankind could easily outlast Earth's overheated demise by moving onward to other solar systems. But can we outlast the stars themselves? Can we somehow circumvent proton decay? Can we manage without the energy-giving properties of black holes? Can life of any kind survive in the final enveloping desolation of the Dark Era?

In the course of this book, we address the prospects and possibilities for life during each successive stage of the universe's future evolution. This analysis is necessarily accompanied by an atmosphere of some uncertainty. A general theoretical understanding of life is quite conspicuous through its absence. Even in the one environment where we have direct experience, our home planet Earth, the ascent of life is not understood. In our brash discussions of the possibilities for life in the far future, we are thus on a qualitatively different footing than when we deal with purely astrophysical phenomena.

Although we have no sound theoretical paradigm that describes how life arises, we need some kind of working model to organize our evaluation of life's prospects for survival and propagation. In order to at least partially cover the range of possibilities, we base our speculations on two very different models for life. In the first and most concretely obvious case, we consider life based on a biochemistry that is roughly similar to that on Earth. Life of this sort will presumably arise on terrestrial planets or perhaps on large moons in other solar systems. In keeping with a time-honored tradition among exobiologists, we assume that as long as liquid water is present on a planet, then carbon-based life can evolve and thrive. The requirement that water must be in liquid form places a rather severe temperature constraint on any potential environment. For example, for atmospheric pressures, the temperature must be larger than 273 degrees kelvin, the freezing point of water, and less than 373 degrees kelvin, the boiling point of water. This range of temperatures excludes most astrophysical environments.

The second class of life-forms is based on a much more abstract model. In this latter case, we draw heavily on the ideas of Freeman Dyson, an influential physicist who has put forth a scaling hypothesis for abstract life-forms. The underlying idea is that at any temperature, one can imagine some kind of abstract life-form which thrives at that temperature—at least in principle. Furthermore, the rate at which this abstract creature uses energy is in direct proportion to its temperature. For example, if we imagine a Dyson organism living at a set temperature, then according to the scaling law, another qualitatively similar life form that happens to thrive at half the temperature would have all of its vital functions slowed down by this same factor of two. In particular, if the Dyson organisms in question are intelligent, and have some form of consciousness, then the effective rate at which they experience events is not given by the real physical time, but rather by a scaled time which is proportional to temperature. In other words, the rate of consciousness for a Dyson organism operating at a low temperature is slower than that of an otherwise comparable life-form operating at a higher temperature.

This abstract approach moves the discussion far beyond the familiar carbon-based life of our planet, but nevertheless it makes certain assumptions about the nature of life in general. Most importantly, we must assume that the ultimate basis for consciousness lies in the *structure* of the life-form and not in the *matter* that makes it up. For instance, in human beings, consciousness somehow arises through a series of complex biochemical processes operating within the brain. The question is whether or not this organic architecture is necessary. If we could somehow build another copy of the entire construction—the person—using a different set of building materials, would the copy be able to think in the same way? Would the copy think it was the same person? If the organic construction turns out to be necessary for some reason, then the *matter* out of which life is made plays a key role, and the possibility for abstract life-forms in widely varying environments is severely limited. If, on the other hand, as we assume here, only the *structure* is necessary, then a multitude of life-forms can exist in a wide variety of different environments. The Dyson scaling hypothesis gives us a rough idea about the metabolic rates and thought rates of these abstract life-forms. This framework is an optimistic one, but as we shall see, it leads to rich and interesting consequences.

A COPERNICAN TIME PRINCIPLE

As our story proceeds and the grand eras of time unfold, the character of the physical universe changes almost completely. A direct consequence of this change is that the universe of the distant future, or the distant past, never resembles the universe that we live in today. Since the present-day universe is rather convenient for life as we know it—we have stars to provide energy and planets to live on—we have a natural tendency to think of the present epoch as privileged in some manner. Resisting this tendency, we adopt the idea of a *Copernican time principle,* which states quite simply that the current cosmological epoch has no special place in time. In other words, interesting things will continue to happen as the universe evolves and changes. Although the available levels of both energy and entropy production become increasingly lower, this effect is compensated by the increasingly long time scales available in the future. Stating this idea yet another way, we claim that the laws of physics do not predict that the universe ever reaches a final quiescent state, but rather that interesting physical processes continue to operate as far into the future as we dare to imagine.

The idea of a Copernican time principle forms a natural extension of our ever widening view of the universe. A major revolution in perspective took place in the 16th century, when Nicolaus Copernicus argued that Earth is not the center of our solar system, as had been previously assumed. Copernicus correctly realized that Earth is just one of a number of planets orbiting the Sun. This apparent reduction in status of Earth and hence mankind had profound repercussions at the time. As the story is usually told, the heretical implications of this shift in thinking prompted Copernicus to delay the publication of his greatest work *De Revolutionibus Orbium Coelestium* until 1543, the year of his death. He hesitated until the very end, and came close to leaving his work concealed. In the introduction of his book, Copernicus writes, "I was almost impelled to put the finished work wholly aside, through the scorn I had reason to anticipate on account of the newness and apparent contrariness of my theory to reason." In spite of the delay, the work was eventually published, and the first printed copy was laid out on the deathbed of Copernicus. Earth was no longer thought to be the center of the universe. A major revolution had begun.

After the Copernican revolution, the reduction in our status not only continued, but accelerated. Astronomers soon realized that other stars are actually objects like our Sun and can, at least in principle, have planetary systems of their own. One of the first to make this inference was Giordano Bruno, who claimed not only that other stars had planets, but also that the planets were inhabited! He was subsequently burned at the stake by the Roman Catholic Inquisition in 1601, although allegedly not for his astronomical assertions. Since that time, the idea that planets might exist in other solar systems has been taken up time and again by eminent scientists, including Leonhard Euler, Immanuel Kant, and Pierre Simon Laplace.

Remarkably, the idea of planets outside our solar system remained a purely theoretical concept, without supporting data, for nearly four centuries. Only in the last few years, beginning in 1995, has the existence of planets orbiting other stars finally been firmly established. With new observational capabilities and a great deal of hard work, Geoff Marcy, Michel Mayor, and their collaborators have shown that planetary systems are relatively commonplace. Our solar system has now been reduced to being but one of perhaps billions of solar systems in the galaxy. Another revolution has begun.

Stepping up to the next larger size scale, we find that our galaxy is not the only galaxy in the universe. As cosmologists first realized in the early part of the 20th century, the visible universe is teeming with galaxies, each containing billions of stars with the potential of having their own planetary systems. Furthermore, just as Copernicus showed that our planet has no special place within our solar system, modern cosmology has shown that our galaxy has no special place within the universe. In fact, the universe seems to obey a *cosmological principle* (see the following chapter), which states that on large scales the universe is the same everywhere in space (the universe is homogeneous) and that the universe looks the same in all directions (the universe is isotropic). The cosmos has no preferred locations nor directions. The universe exhibits an astonishing regularity and simplicity.

Each successive demotion of Earth's central status leads to the irrevocable conclusion that our planet has no special location within the entire universe. Earth is an ordinary planet orbiting a moderately bright star in an unexceptional galaxy located at a random position within the universe. The Copernican time principle extends this general idea from the spatial to the temporal domain. Just as our planet, and hence mankind, has no special location within the universe, our current cosmological epoch has no special place in the vast expanses of time. This principle further erodes the remaining vestiges of anthropocentric thought.

This book is being written near the close of the 20th century, an appropriate time to reflect upon our place in the universe. Using the breadth of understanding gained during this century, we can take an unprecedented look at our position in both time and space. In accordance with the Copernican time principle and the rich variety of astrophysical events yet to take place in the vast expanses of future time, we argue that as this millennium draws to a close, the end of the universe is not very near. Armed with the four forces of nature, four astronomical windows to view the universe, and a new calendar that measures time in cosmological decades, we now set forth on our journey across the five grand eras of time.

1

THE PRIMORDIAL ERA

$-50 < \eta < 5$

A VIOLENT EXPLOSION LAUNCHES THE UNIVERSE ON ITS EVOLUTIONARY VOYAGE INTO THE FUTURE.

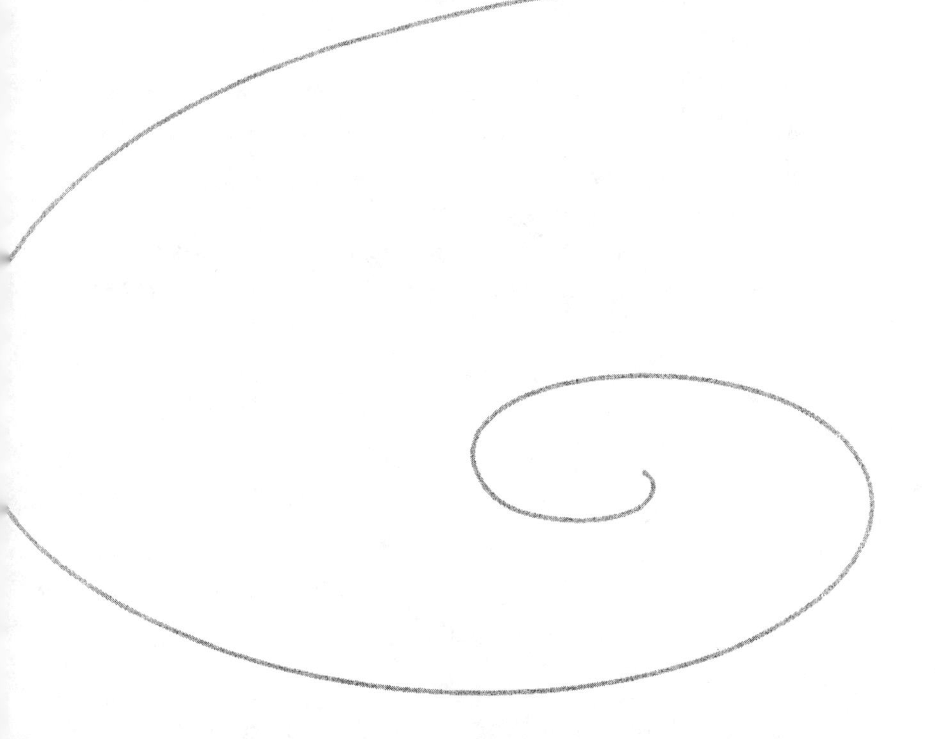

Before the beginning, beyond the horizon:

Chaos rules. The underlying fabric of space-time is going crazy. Instead of exhibiting the usual three dimensions of ordinary space, with a natural and directed flow of time, space-time is a roiling, random, and fluctuating froth that continually changes its geometry every 10^{-43} seconds. No clean separation of space and time is meaningful in this ultra-high energy regime of physical reality. Quantum mechanics and gravity engage in a cosmic battle with universal implications.

Out of this probabilistic foam erupt explosive bubbles of microscopic space-time. Most of these regions are destined to expand and recollapse back into the amorphous foam within incomprehensibly short stretches of time. Every so often, when the conditions are just right, an erupting region of space-time launches itself into a state of fantastically rapid expansion. The ensuing exponential growth is so violent that adjacent points of space race away from each other at rates exceeding the speed of light. This counterintuitive activity increases the size of the expanding region by an extraordinarily large amount within a minuscule time interval.

After this fast-paced epoch of superluminal expansion, the inflated region of space-time is stretched far beyond the original background foam, so that it becomes effectively separated. Once jump-started on this trajectory, the newly isolated region can continue its expansion forever, albeit at a slower rate. The region grows to unfathomably large sizes as it endures the unending future vistas of time. The newly born universe thus begins its long and tortuous journey towards death.

A major achievement of modern science is the establishment of a theory for the origin of the universe. The universe began with an explosive event at a well defined point in time, and expanded to become our universe of today. With the advent of the big bang theory, many questions once consigned to philosophy or metaphysics regarding the birth of the universe are now understood in stark rigid detail. The moment of creation has been placed on a solid scientific foundation, removed from idle conjecture, and verified in many respects by experiment and observation.

Imagine what it would be like to witness the beginning of time. If we could experience these first defining moments, if our eyes could observe the microscopic events taking place with blinding speed, what would we see? Let's pick up the story just as the universe bursts into existence. We would first notice that the universe is expanding and cooling at a fantastic rate. During the first 10^{-35} seconds or so, the universe expands so fast that adjacent points of space rush away from each other at incomprehensible speeds. During this brief period of counterintuitive behavior, a portion of the universe the size of a small dot (·) inflates to become larger than the entire observable universe of today. The expansion soon slows down to more reasonable subluminal speeds, but continues onward nonetheless.

We would also notice that the early universe was very hot and very bright. At such enormous temperatures, much hotter than the central regions of any star, matter is vastly different from the material of everyday experience. On Earth, ordinary matter is made up of atoms, each composed of a set of electrons orbiting a nucleus containing protons and neutrons. In the extreme heat of the first microseconds, the temperature is too hot for molecules, atoms, and nuclei to be bound together. Even protons and neutrons cannot exist. The universe is swarming with mysterious elementary particles called quarks.

Under ordinary circumstances, we think that matter makes up everything in the universe. Right now, for example, a large portion of the mass-energy in the universe is contained in the matter within galaxies, with very little in between. During the earliest moments of history, however, when matter was broken down into its basic particle constituents, the universe had a very peculiar aspect. The particles of matter constituted only a tiny fraction of the total energy density of the universe. Most of the energy density was contained in the background field of radiation, and the universe resembled an extraordinarily hot oven, a primordial blast furnace.

The radiation field that was present at the beginning is still with us today. It forms a sea of photons that fills all of space and is called the *cosmic background radiation.* This radiation background now contains much less energy than it held in the distant past. Its effective temperature has fallen to a frigid 2.7 degrees above absolute zero. At early times, however, the background radiation was exceedingly bright and hot. The expansion of the universe has subsequently stretched the once intense background light into millimeter-long microwaves. The blast furnace of the past has been degraded into a low-energy microwave oven.

When the universe is one microsecond old, we are immersed in a vast sea of radiation, with a relatively small admixture of quarks and other particles. The quarks are made of both ordinary matter and antimatter, with a slight excess of the former. For every thirty million antimatter quarks in its storehouse, the universe contains thirty million and one quarks made of matter. As the universe evolves and cools, the quarks and the antiquarks annihilate with one another. Only the tiny excess fraction of matter survives the process. This seemingly insignificant residue eventually makes up all the matter that we see in the universe today—the galaxies, the stars, the planets, you and me.

As the quarks and antiquarks annihilate, the leftover quarks begin to condense into protons and neutrons. After about thirty microseconds, no free quarks are left. Because the universe is still completely dominated by radiation (photons) rather than matter particles, the universe itself is hardly troubled by this change in its particle inventory, and the expansion continues relentlessly.

Next, as the universe continues to cool, protons and neutrons begin fusing into helium and other light nuclei. This process starts when the universe is about one second old and ends rather abruptly after a few minutes. Most of the helium that exists today was forged during this early burst of nuclear activity. Heavier nuclei like carbon and oxygen, the elements that provide the raw material of life, do not form yet. The universe expands much too quickly for large nuclei to come together. The available time is too short and the density of the universe is too low. The heavy elements are produced much later in dense stellar cores and during supernova explosions that mark the deaths of massive stars.

With the production of the light elements, the particle content of the universe undergoes a substantial change. This restocking of the particle inventory is the second such event during the first few minutes of time. Radiation continues to dominate the universe as its main constituent. And the expansion continues.

INFLATION

As we have already noted, in the very earliest moments of cosmic history, the universe was caught in a brief but intense phase of incredibly rapid expansion. This period of superluminal expansion inflated the size of the universe by an enormous factor, perhaps a million trillion trillion (10^{30}), and probably much more. After this short but universe-altering epoch, the cosmos settled down into a state of more ordinary expansion. How and why did this inflation come about?

Conventional big bang theory accounts for the expansion of the universe, the abundances of the light elements, and the existence of a cosmic background radiation field. The theory has the additional advantage of being both mathematically simple and elegant. In its original form, however, the big bang theory does not provide a complete explanation of the universe. Fortunately, many of the remaining properties of our universe—namely its large size, its flatness, and its extreme uniformity—can be accounted for by a single modification. This extension to the theory, known as *inflation,* was put forth by Alan Guth, now a professor at M.I.T. His seminal work—*The Inflationary Universe,* published in 1981—sparked a revolution in cosmological research.

The process of inflation naturally explains why the universe is so large and so uniform. Inflation also flattens out the geometry of space-time to the degree we now observe in the cosmos. The basic idea of inflation is simple: early in the history of the universe, it suddenly expanded in size by a huge amount. In order to evolve into a universe resembling our own, with the properties we see today, the primordial universe had to expand by a factor of at least 10^{28}. To get some feeling for the immensity of this number, notice that the size of the currently observable universe is about 10^{28} centimeters. So inflation is like blowing up a pebble to the size of our entire observable universe, or larger, all within a tiny fraction of a second. This extremely rapid expansion takes place if the energy density of the universe is dominated by *vacuum energy density.* This rather mysterious type of energy density exhibits the curious property of a negative pressure. If the energy density of the universe is dominated by vacuum energy, the negative pressure will drive an ever increasing rate of expansion. This accelerating expansion can inflate the universe by the large amount required to explain the properties of our universe.

At first glance, the concept of a vacuum energy density seems like a contradiction in terms. We usually think of a vacuum as being completely empty. How can something that is supposedly empty have any energy at all, let alone dominate the energy density of the entire universe? At the fundamental level, the vacuum must obey a quantum mechanical description, which means that *the vacuum is not really empty after all.* The vacuum is governed by the Heisenberg uncertainty principle, named after Werner Heisenberg, a pioneer of quantum mechanics. This fundamental concept in quantum mechanics arises because of the wavy nature of physical reality at small size scales and leads to the possibility of vacuum energy.

Consider an electron for example. Heisenberg's uncertainty principle states that we cannot simultaneously measure both the particle's momentum and the particle's position to an arbitrarily high degree of accuracy. Since we cannot measure both the momentum and the position precisely, the uncertainties in the momentum and the position cannot be made too small at the same time. In other words, the joint uncertainties must be larger than some number, usually denoted as $\hbar$. The quantity $\hbar$ is a fundamental constant of nature known as *Planck's constant.* An analogous law says that the uncertainty in energy and the uncertainty in time cannot both be made too small at once. An important consequence of this uncertainty principle is that nature can violate conservation of energy, provided that the criminal act of nonconservation occurs for a sufficiently short time.

The constant $\hbar$ is almost vanishingly small when viewed from the perspective of ordinary experience. As you watch a car driving down the street, you experience no problem in simultaneously knowing both the position and velocity of the car. The need for accuracy in ordinary measurements (about one inch to determine the car's position, and about one mile per hour to determine the speed) is entirely uncompromised by the intrinsic uncertainty associated with Heisenberg's principle.

The uncertainty principle has important consequences for the concept of a vacuum. From a quantum viewpoint, the vacuum cannot really be empty. Imagine a region of space devoid of matter, a region we would normally consider to be "empty." Because of the uncertainty principle, this apparently empty space is filled with particles flickering briefly in and out of existence. The energy required to make these particles is borrowed from the vacuum and then quickly repaid when the particles annihilate each other and subsequently disappear back into nothingness. These particles are called *virtual particles* because they have no real lives. They live on borrowed time and always annihilate just after their spontaneous appearance out of the vacuum. Because of the spontaneous creation of these virtual particles, otherwise empty space is seething with these ghostly entities. And these virtual particles can endow the vacuum with an effective energy density that it would not otherwise possess. Quantum behavior thus leads naturally to the concept of empty space being allowed to have an energy density.

Not only can the vacuum have an associated energy density, but it can also attain different energy states. A rather high-energy vacuum state is required for the universe to inflate during the earliest moments of time. But this vacuum energy is considerably larger than the vacuum energy state of today's universe. At the present time in history, the vacuum does not play a dominant role in the dynamics of the universe; otherwise, the universe would be expanding quite differently than it is today. Thus, in order for the inflationary universe scenario to work, the vacuum energy density of the universe must be incredibly large at some early time in history, and must now be very small (or zero). If the vacuum energy is not precisely zero at the present time, however, it will have dramatic consequences for the future, as we shall see later on.

In the early universe, when the background temperature is high enough so that the strong, weak, and electromagnetic forces are unified, the vacuum energy density can dominate the universe. When such vacuum domination occurs, the universe enters its phase of inflation and expands rapidly. If the resulting inflationary phase of expansion lasts long enough, the universe can expand by the magic factor of 10^{28}, the amount required to produce our universe of today, and perhaps much more.

The universe can become dominated by vacuum energy in many different ways. Many particle theories suggest that nature contains entities known as *scalar fields.* These quantum mechanical fields are somewhat analogous to the electric potential that gives rise to the more familiar electric force. Scalar fields, however, can have intrinsically large energy scales, far larger than the energies explored by present-day particle accelerators. As a result, these scalar fields remain a purely theoretical construct; no direct experiments have yet been made to verify their existence. The potential energies associated with the scalar fields can make dramatically large contributions to the vacuum energy, large enough to dominate the energy density of the universe at very early times. For example, at the energy density associated with grand unification of the forces, a cubic centimeter of the vacuum would contain more energy than the entire observable universe today!

The large energy associated with the vacuum greatly affects the expansion of the universe. As Einstein's famous formula $E = mc^2$ illustrates, energy is equivalent to mass, and these large vacuum energies must have correspondingly large gravitational effects. A large amount of energy implies a large amount of mass, which pulls matter together and usually slows down the expansion rather than accelerating it. The vacuum energy, however, has the curious property of exhibiting a *negative pressure*. This negative pressure is larger than the mass energy and hence the pressure effects dominate the expansion. Even though we usually don't think of pressure having a gravitational effect, the theory of general relativity demands that it does. Ordinary positive pressure pushes outward, and its gravitational effect pulls inward. On the other hand, the gravity of a negative pressure pushes outward, exactly the behavior that drives the universe to expand at an accelerating rate. The net result of this stupendously large and negative pressure is to enormously enlarge the universe within a tiny fraction of a second. As this inflation proceeds to completion, the cosmos attains its properties of flatness and uniformity that we observe today.

THE HORIZON AND FLATNESS PROBLEMS

One important property of our universe is that it looks the same in all directions. In particular, the temperature of the cosmic background radiation is nearly identical in different directions in the sky. This radiation was emitted by different regions of the universe, which must have communicated with each other in order to have the same temperature. And this communication must have taken place before the universe was 300,000 years old, when this radiation last interacted. Without inflation, such regions are not able to communicate with each other and the universe suffers from the *horizon problem*. The inflationary universe elegantly circumvents this difficulty, as we shall see.

As the universe expands, two effects occur simultaneously. First, the universe itself expands, which simply means that the space-time of the universe grows larger. Second, the universe grows older so that light signals have a longer time to travel and hence larger regions of the universe can become in *causal contact*. If some event that happens at one location and time can influence something at a different point in space and time, then the two events are said to be in causal contact. For example, you can influence the events that will occur a minute from now in the room where you are reading this book—perhaps by lighting a fire and burning down the room. Nothing you might conceivably do, however, can possibly influence anything that will happen on the planet Mars during the next minute. Mars is more than one light minute away, and no signal that carries information can travel faster than the speed of light.

When not inflating, the universe expands at a rate "less than the speed of light." On the other hand, because its effective size is determined by the distance that light signals can travel, the portion of the universe that is allowed to be in causal contact grows at the speed of light. Taken together, these two results imply that the universe contained within the causal horizon is growing with time. In other words, as time marches onward, new material is continually becoming part of what we consider to be the observable universe. The time scale on which the universe changes in this manner is now billions of years, roughly comparable to the current age of the universe. As a result, although the universe continues to grow, such changes go unnoticed over the course of a human lifespan.

We can now state the horizon problem more precisely. When we observe the cosmic background radiation, we are effectively looking back in time to when the universe was about 300,000 years old. The background radiation was last able to interact with matter during this epoch and the background photons we see today have been streaming freely since that time. The size of the speed-of-light sphere, which sets the causality boundary, was thus only 300,000 light-years across when the cosmic background radiation was emitted. Because the universe has expanded since that time, these regions have expanded to a size of about 300 million light-years at the present epoch. When we observe the cosmic background radiation by looking in opposite directions in the sky, however, we are sampling regions separated by the size of the entire observable universe today, distances larger than 20 billion light-years. This distance is far greater than the size of the regions that are allowed to be in causal contact, yet the observed temperatures of the cosmic background radiation are virtually *identical,* the same to a few parts in 100,000. From *a priori* considerations, there is no reason why the temperatures of regions which were completely out of contact should be so uniform. This dilemma constitutes the horizon problem.

This horizon problem is naturally resolved by inflation. Imagine a tiny portion of the universe that is in causal contact with itself at the moment before the inflationary period of expansion begins. By definition, such a region must be smaller in size than the speed of light times the age of the universe at that time. Now let that small region inflate its size by an absolutely stupendous amount. If the growth factor is large enough, then the entire observable universe that we see today can be contained within the region of causal contact that we started with. The required value of the growth factor is the same magical factor of 10^{28} that we met before. In a universe that experiences inflation, the sizes of regions that have already been in causal contact, the regions with the possibility of having the same characteristics, are much larger than those in a noninflating universe (see the figure below).

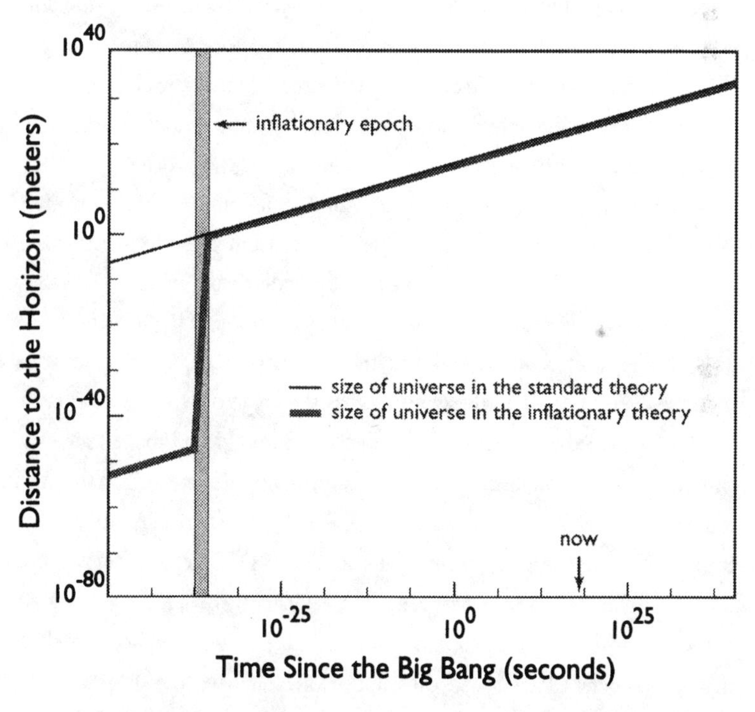

This figure shows the size of the universe in both the standard big bang theory and in the modified version incorporating an inflationary phase of expansion. The inflationary model is a solution to the horizon problem because it allows for a much smaller universe in the very distant past. This smaller universe could have been in causal contact with itself at some early time in history and thereby produce the extreme uniformity that we observe in our universe today.

Another problem facing a cosmology without inflation is commonly known as the *flatness problem.* The problem here is that we observe the spatial geometry of the universe to be very flat, which means that the density of the universe is rather close to a certain critical value. A flat universe has exactly this critical density and is destined to expand forever, but at a continually diminishing rate. In order for the universe to have this property today, the initial conditions for the expansion of the universe had to be very special and hence very unlikely to occur.

An expanding universe can be open, closed, or flat, where a flat universe is the intermediate case in which the universe expands forever but barely so. When we make measurements to determine the amount of matter in our universe, we find that the universe is close to being flat. The density of the universe has the critical value—the density required for the universe to be flat—to within a factor of two or three. More precisely, the ratio Ω_0 of the total energy density of the universe to the critical value appears to lie in the range $0.3 < \Omega_0 < 2$, a range that includes the flat case $\Omega_0 = 1$.

So, how does this constitute a problem? The difficulty arises because the ratio Ω, a measure of how far the universe is from the critical flat case, changes with time. If the universe is a little bit underdense (which means that $\Omega < 1$) at a given cosmological time, then the expansion of the universe wins its battle with gravity. As time goes on, the ensuing rapid expansion makes the universe even more underdense and forces the value of Ω to become even smaller. Thus, if the universe is underdense at some given time, a relatively short time later the value of Ω becomes extremely small, far less than the critical value of unity. In order to have the value of Ω anywhere close to unity at the present time, the universe must have a value of Ω in the past that was extraordinarily close to the critical value of 1, but slightly less. If we go back all the way to when the initial conditions for the universe were set up, when less than 10^{-43} seconds had elapsed, the value of Ω would have to be frighteningly close to unity, with an enormous accuracy of about one part in 10^{60}. It is difficult to understand why the universe would produce the inordinately precise value of the energy density required to have Ω lie close to unity today. Similarly, if the universe is a little bit overdense, gravity will win its battle with the expansion, and the ratio Ω will rapidly grow much larger than unity.

But the cosmological flatness problem is also alleviated by a colossal inflationary expansion of the universe. To illustrate the resolution of the flatness problem, let's inflate a balloon and use its surface as a two-dimensional model of the universe. The surface of a balloon is curved and this curvature represents the curvature of space-time. If we blow up the balloon to some grossly distended size, so that its radius is, say, 10^{28} times larger, then the surface of the balloon will seem much flatter. If we start with an ordinary balloon, with a radius of about 10 centimeters, then the final size of the inflated balloon will be larger than the entire observable universe at the present time. Just as the surface of the balloon becomes flatter under the action of extreme expansion, the curvature of space-time becomes flattened out as the universe blows up its size by an enormous amount (see the figure on the next page).

THE EXPANDING UNIVERSE

In the 1920s, Edwin Hubble used the 100-inch telescope on Mount Wilson to demonstrate that the universe is expanding. This discovery was nothing short of monumental. Before Hubble's breakthrough, which was actually an extension of slightly earlier work by Vesto Slipher, scientists had generally assumed that the universe is static and unchanging. The realization that we live in an expanding universe greatly transformed our cosmic perspective.

Hubble observed that all but the very closest galaxies are receding from our galaxy—the Milky Way. Hubble also showed that the farther away a galaxy is, the faster it moves away from us. This relationship between distance and velocity is now called *Hubble's law* and is a natural consequence of an expanding universe as viewed from within.

We can consider a simple model of the universe to illustrate the Hubble expansion. Imagine a large raisin cake, with the embedded raisins representing the galaxies in this model. As the cake expands, while it bakes, all of the raisins in the cake move away from all of the other raisins. The universe does not have any kind of edge, however, at least not that we know about. As a result, we must think of the universe as a very large raisin cake, with our raisin—the Milky Way galaxy—far away from any kind of boundary.

The expansion and evolution of the universe is described by Einstein's

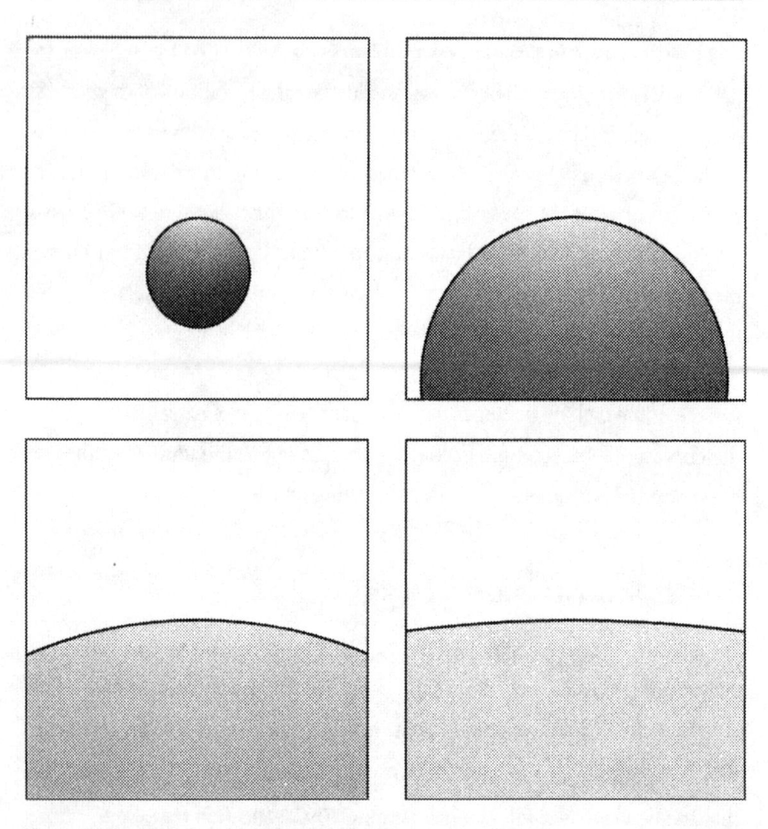

These four panels show the surface of a sphere, where the curvature of the surfaces represents the curvature of the space-time of the universe. The radius of the sphere grows three times larger between each successive snapshot. The first spherical surface is small and clearly curved, whereas the fourth and final surface is difficult to distinguish from a plane. The expansion of the universe during inflation flattens out the universe in a similar way, and hence provides a solution to the flatness problem. During inflation, however, the universe grows more than 10^{28} times larger, entirely swamping the meager 27-fold increase shown here.

theory of general relativity, which incorporates gravity into a fundamental description of space-time. When general relativity is used to describe the universe as a whole, cosmic expansion becomes a natural consequence of the theory. When Einstein realized that his theory implied an expanding universe, he initially assumed that the prediction was incorrect. Astronomers had not yet observed the Hubble expansion, and almost everyone still subscribed to the antediluvian view of a static and unchanging universe. Einstein went so far as to unnecessarily complicate his theory to allow for a static (not expanding) universe. After the expansion was discovered, however, cosmologists quickly understood that the original, unmodified equations provided the best description of our continually growing universe.

The universe obeys a basic tenet called *the cosmological principle*, a guiding postulate that organizes and simplifies the possible behavior of the cosmos. This principle asserts that the universe as a whole is both *homogeneous* and *isotropic*. A homogeneous universe is one that is the same at all points in space; in other words, the universe has no uniquely preferred location (see the figure on the right on page 16). Similarly, an isotropic universe is one that looks the same in all directions; in other words, the universe does not look different when you look out in different directions from our galaxy (see the figure on the left on page 16). The cosmological principle is a natural generalization of the Copernican viewpoint. Copernicus showed that Earth and hence mankind do not occupy a special location in our solar system. Because the universe is homogeneous and isotropic, our galaxy does not lie at a special location. In particular, we do not live at the center of the universe.

Our expanding universe is not infinitely old, but rather has a definite age. If we start with the expansion of the universe as observed today, and then "run the clock backward" to extrapolate the motion to earlier times, all of the matter in the universe reaches an infinite density at a given time in the past. This singular point represents the big bang itself, which defines the beginning of time. The span of time starting at this point and ending at the present epoch is the current age of the universe, about ten billion years.

POSSIBLE FATES FOR AN EXPANDING UNIVERSE

An expanding universe has at least three possible long-term prospects. First, if the expansion continues unabated forever, the universe is said to be *open*. Alternatively, if the universe is destined to eventually halt its expansion and

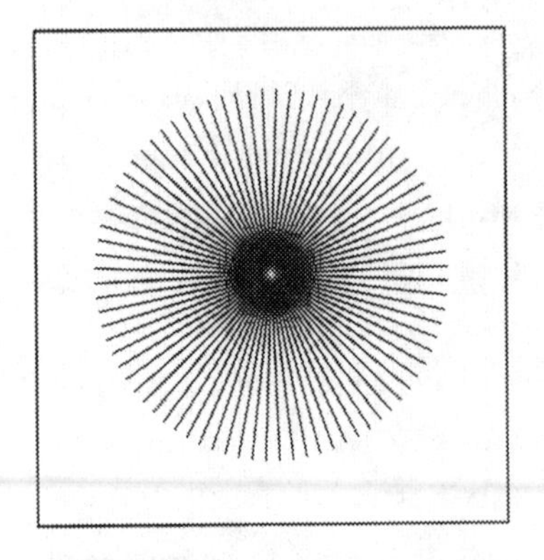

Because all the sharp points on this object emanate from a central point, this system (left panel) is *isotropic* but not *homogeneous*. The central point enjoys a special location within the system. The aggregate of sharp points does not, however, display any preferred direction or alignments. Because none of the leaves in this pattern has a special location, the arrangement (right panel) is *homogeneous*. However, the pattern has two preferred directions along which the leaves tend to align themselves and which reflect the underlying geometry of the arrangement. With these preferred directions present, it cannot be *isotropic*.

recollapse, then it is said to be *closed*. A *flat* universe lies at the point of division between the open and closed possibilities. In a flat universe, expansion continues forever, but at a continually slower rate. As the age of the universe becomes infinite, the expansion gradually slows to a complete stop.

The long-term fate of the universe hinges dramatically on whether the universe is open, closed, or flat. These choices are determined by the amount of energy density in the universe. The best available astronomical evidence suggests that our universe does not contain enough energy density to be closed, and hence the cosmos should expand forever. If the universe is indeed open, or flat, then it will live long enough for a fascinating series of dramas to develop. In contrast, a closed universe would keep the breadth of possibilities on a much shorter leash.

For now, we can understand the three different types of universes using a simple analogy. Imagine launching a rocket ship from the surface of a planet. The rocket soars straight up with an initial burst of speed, and then switches off its engine. What happens? The answer depends on how fast the rocket is going, or, looking at the problem another way, on the total mass of the planet and the rocket. If the rocket is too slow, or if the planet has too much mass, then the rocket cannot escape the gravitational field and falls back to the surface. We can think of the rocket and the planet as a *closed* physical system, in which the rocket and planet fall back toward each other, or recollapse. This same set of circumstances arises when we throw a baseball and it falls back to Earth. On the other hand, if the rocket in our example departs with sufficient speed, it can escape the gravity of the planet and continue its travels forever. This situation corresponds to an *open* physical system, like an open universe that expands indefinitely.

Both the universe and our simple planet-rocket system have an important intermediate case that we label as *flat*. The rocket might be launched with a speed that gives it exactly enough energy to escape from the planet. As it travels, the rocket continues to slow down and eventually comes to a complete stop when it reaches a point in space infinitely far away from the planet. Of course, the rocket needs an infinite amount of time to reach this point. A flat universe behaves in a qualitatively similar manner. All of the galaxies expand away from each other, but they slow down with time. The galaxies approach the static state—a complete stop—as the universe becomes infinitely old.

Although this model provides a convenient framework to discuss open, flat, and closed universes, an important distinction remains. The rocket and the planet cruise through space in the way that we normally visualize travel. Within the universe, however, space itself is expanding, and the galaxies are really locally at rest. The rocket and the planet constitute a *classical* system, whereas an expanding universe is an example of *general relativistic space-time*.

THE COSMIC BACKGROUND RADIATION

The entire universe is pervaded by a background sea of radiation. If the universe is expanding, then at earlier times it must have been smaller, denser, and hotter. At extremely high temperatures, particles and radiation exist in a kind of equilibrium. During these hot dense early phases, a great deal of radiation was present. As the universe expands and cools, the radiation is stretched out to lower energy and eventually ceases to interact. Some of this radiation is left over and is streaming freely through space today in the form of microwaves.

Although no longer consequential, this background radiation remains detectable. It provides an unmistakable signature of a blazing distant past.

This microwave background radiation was discovered in 1965 by Arno Penzias and Robert Wilson at the AT&T Bell Laboratories. Penzias and Wilson were making a careful survey of background sources of static—in other words, white noise—and were not expecting to find the skies filled with low energy microwaves. Their serendipitous discovery earned them the Nobel Prize in physics.

How do we know that this faint background of microwaves is actually a fossil signature of the big bang? After all, many other physical processes generate radiation. Many people are worried about radiation from nuclear power plants. Television and radio stations are constantly disgorging low-energy radiation into space. And on a larger scale, stars are continually pumping copious quantities of radiation into the galaxy.

Early in cosmic history, the universe was dense and hot. Under these conditions, which are vastly different from the sparse and cold intergalactic voids of today, the radiation field which permeated all of space reached a state of thermodynamic equilibrium. When equilibrium is attained, the spectrum of the radiation, the amount of energy emitted at each wavelength, approaches a particular form known as a *blackbody*. This same spectral distribution of wavelengths is emitted by any perfectly black object (that is, opaque and nonreflecting) in thermal equilibrium at a specific temperature. Every blackbody spectrum of radiation corresponds to one particular temperature. For example, the Sun emits a nearly blackbody spectrum from its surface, with a temperature of 5800 degrees kelvin. The background radiation of the universe also has a particular temperature. As the universe expands and cools, this characteristic temperature decreases, but the distribution of radiation retains its distinctive shape, its overall blackbody form.

The cosmic background radiation, as measured today, has a temperature of 2.73 degrees kelvin. Furthermore, the radiation spectrum has almost a perfect blackbody shape, accurate to about one part in ten thousand. The best measurement of this spectrum was made by the COsmic Background Explorer (COBE) satellite in the late 1980s (see the figure opposite). This finding provides a dramatic endorsement of the big bang theory.

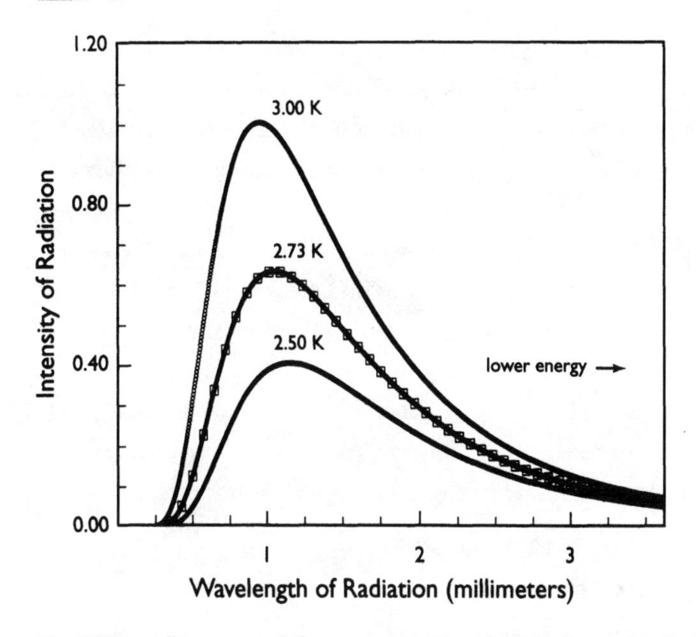

The COBE satellite measured the spectrum of the cosmic background radiation and each small square here represents an individual measurement. The three curves plot the intensity of blackbody radiation, versus wavelength, for three different temperatures. Notice how closely the COBE measurements match the 2.73 degree blackbody curve. The big bang theory predicts this particular form of the curve.

Another property of the observed background radiation was imprinted in the blazing past of the early universe. The cosmic background radiation looks the same in all directions. Almost. As described above, the radiation field has a blackbody form and can be characterized by a single temperature. And this temperature is nearly the same in all parts of the sky to an extremely high accuracy. This result is in keeping with the cosmological principle, which holds that the universe is homogeneous and isotropic.

Nevertheless, tiny fluctuations in the temperature of the cosmic background radiation have been observed. These fluctuations have amplitudes of only about 30 parts per million and were first detected by the COBE satellite in 1992. These minute fluctuations have profound consequences. The radiation background last interacted with matter when the universe was 300,000 years old. This well defined turning point, known as *recombination*, corresponds to the time when the universe became cool enough for electrons and nuclei to form atoms. (From a logical standpoint, we should speak of a "first combination" rather than a "recombination" because the electrons and the nuclei were never previously bound together.) Before recombination, the radiation and the matter in the universe interacted strongly and were closely coupled. After atoms appeared, however, the universe suddenly became transparent to its background radiation. The fluctuations observed in today's cosmic background are an imprint of the matter density fluctuations which existed when the radiation and matter were last in contact. Because the fluctuations in the matter density ultimately grew into galaxies and galaxy clusters, the fluctuations in the microwave background are the signature of the initial conditions for galaxy formation and larger scale structures.

QUARKS AND ANTIQUARKS

During the first microsecond of cosmic history, the matter content of the universe lived in the form of quarks and their antimatter counterparts called antiquarks. These somewhat mysterious particles are components of the more familiar protons and neutrons that make up most of the matter we know today. At high temperatures, however, matter prefers to live in the form of free quarks rather than in larger entities like protons. Although most are destined for annihilation, some fraction of these primordial quarks survive to eventually make up the matter of today's universe. But long before protons and neutrons even existed, microscopic events of great import took place to shape the future matter inventory of the universe.

Our universe today is primarily composed of matter, rather than antimatter. When placed sufficiently close together, matter and antimatter annihilate with one another and leave behind an energetic burst of radiation. Essentially all of the mass is converted into energy by this process. But as we walk around on our planet, we never experience such annihilation events. Why not? Because Earth is almost exclusively made of matter and not antimatter. The fact that NASA missions landing on the Moon, and then Mars, have not ended in dramatic explosions of radiation strongly indicates that our solar system is also composed of matter, with virtually no antimatter. Observing larger

The Darkness of the Night Sky

The finite age accorded to the universe by modern cosmology solves a classical problem: "Why is the sky dark at night?" Johannes Keppler was probably the first to grasp the importance of this question in the 1600s, although it wasn't until the work of H. W. M. Olbers in the 19th century that this problem became widely known. In 1823, Olbers, a German astronomer, presented a paper that introduced this issue, which subsequently became known as *Olbers' paradox.*

At first glance the answer seems obvious—of course the night sky is dark! After all, the Sun is not lighting up the sky at night. On deeper reflection, however, the problem grows more disturbing. Let's consider the portrait of the universe from the 19th century: a static, infinite universe displaying the ordinary three-dimensional space of Euclidean geometry. Now imagine looking out at the night sky. As you follow a line of sight in *any* direction, you must eventually intercept the surface of a star. And stars are bright. The night sky should be ablaze with the radiative intensity of a stellar surface, as bright as the Sun.

We can think about this issue another way. In this antiquated model of the universe, the sky is littered with an infinite number of stars. The farther away the stars lie from Earth, the dimmer they appear to us. The stars get dimmer by the square of the distance (r^2) from the observer. The volume of the universe and hence the total number of stars, however, grows larger by the cube of the distance (r^3). Even though the stars become dimmer with increasing distance, the effect is compensated by the increasing number of stars. If this model is correct, the night sky should be very bright.

We now know that this old-fashioned paradigm of an infinite static universe is simply wrong. The universe has both a finite age and a non-Euclidean space-time geometry. Because only ten billion years have elapsed thus far, we can only observe stars out to a large, but still strictly finite, distance of ten billion light-years. The observable volume of the universe contains a large but finite number of stars, about one thousand billion billion (10^{21}). These stars contribute to the observed brightness of the night sky, which glows very faintly. The night sky is, however, much darker than the surface of a star.

(continued)

The expansion of the universe also acts to darken the night sky. Because space-time is expanding, the distant stars make less of a contribution to the sky brightness than suggested by the forerunning Euclidean argument. Far-off stars in distant galaxies are moving away from us at speeds approaching the speed of light. This starlight from the furthest reaches of the observable universe is severely stretched out, and thereby diminished in intensity.

The darkness of the night skies has profound consequences for the development and continued existence of life. If the universe had neither a finite age nor an expanding configuration, the night sky really would be as bright as a stellar surface. Under such conditions, stellar evolution would be drastically altered, and the evolution of life on planets would be virtually impossible. If our solar system were transplanted into such a hypothetical bright universe, the Sun and the planets would suddenly find themselves immersed in a thermal bath of radiation with an effective temperature as hot as a star. Because heat must flow from hotter regions to cooler regions, from the second law of thermodynamics, the Sun would heat up in order to radiate its energy into space. The planets themselves would be baked to stellar temperatures, thousands of degrees kelvin, and gradually obliterated by the relentless and energetic flux of background light.

The observed darkness of the night sky strongly suggests a finite age for the universe. This realization is truly remarkable. It is almost equally remarkable that this important clue was overlooked by scientists grappling with Olbers' paradox prior to the 20th century. The idea of a static and unchanging universe was deeply ingrained in the culture. The simple and correct resolution to the paradox was not recognized until Hubble discovered the expansion of the universe, and Einstein wrote down the theory that admitted, and indeed predicted, an expanding space-time.

scales, such as the galaxy and even the entire universe, we also deduce the presence of matter and the marked absence of antimatter. In rough terms, our universe contains about 10^{78} protons and neutrons, with a relatively negligible admixture of antiprotons and other antimatter particles.

In spite of this overwhelming asymmetry observed in our universe, however, the laws of physics do not favor matter over antimatter. According to these basic laws, which have been verified countless times in laboratory experiments, both matter and antimatter are fundamentally on an equal footing. And yet a universal imbalance exists. Clearly something curious is afoot.

Particles made of ordinary matter, like protons and neutrons, are called *baryons.* Particles of antimatter are called *antibaryons.* Our universe thus exhibits a net *baryon number,* which is defined to be the total number of baryons minus the total number of antibaryons. In order for the cosmos to achieve this end result, the laws of physics must allow for a physical process that does not strictly conserve baryon number. The existence of this process—one that violates baryon number conservation—has profound repercussions, both for the genesis of matter in the early universe and for the long-term fate of matter in the far future.

Regarding this latter issue, the failure of the laws of physics to strictly conserve baryon number implies that protons, neutrons, and all ordinary matter are doomed. If we wait long enough, this relatively weak process that violates baryon number *will* eventually drive the decay and destruction of all ordinary matter. Because of the relative inefficiency of the process, however, this part of the story does not become urgent for quite a long time, perhaps a trillion trillion trillion years.

Back in the earliest moments of cosmic history, well before the universe was a microsecond old, physical processes that do not conserve baryon number began to operate. At the high temperatures of this epoch, these processes that violate baryon number are much more effective than at the low temperature of today's universe. The ensuing microscopic reactions produce a net excess of quarks in some regions of the universe, and perhaps an excess of antiquarks in other universes. As our universe expands and cools, the reactions cease to operate, and the relative populations of quarks and antiquarks become frozen in place. The cosmos successfully achieves *baryogenesis,* the production of a net excess of matter over antimatter.

Several different models of this process have been proposed, but they remain under study. Although this discussion is necessarily a little vague, we do understand the baryogenesis process in broad general terms. In order to keep the production mechanism from going the other way, and destroying the excess quarks, the reactions must take place out of equilibrium so that some of the newly forged extra quarks remain intact. The expansion of the universe facilitates the out-of-equilibrium character of the reactions by providing an ever-changing background state. Another condition is also necessary. The microscopic reactions that produce a net baryon number must not be exactly *time reversible,* as are most processes involving elementary particles. The reactions must be able to sense or follow the direction of time, which is defined by the expanding universe. Thus, in order to generate a net excess of matter, the universe must provide reactions that violate baryon number, take place out of equilibrium, and are not time reversible.

The excess baryon number produced in this manner is almost embarrassingly small. For every thirty million antiquarks in existence, the universe contained thirty million and *one* quarks of ordinary matter. This phenomenally small excess—one part in thirty million—is vitally important. As the universe expands and cools, quarks and antiquarks annihilate with each other. Only the extra quarks, those that cannot find an antimatter partner to annihilate with, are left to eventually fill our universe with matter.

When the universe finally becomes cool enough, the quarks condense into composite particles called *hadrons,* which include the protons and neutrons that we know today. This change of phase occurs as the universe falls through a temperature of one trillion degrees kelvin and attains the density of nuclear matter, one quadrillion times the density of water. With these background conditions, protons and neutrons burst into existence. These basic building blocks, synthesized during the first microsecond of history, live not only to the present time, some ten billion years later, but endure into the far future. These particles last for at least ten billion trillion (10^{22}) times longer than the current age of the universe, and perhaps even longer.

NUCLEOSYNTHESIS

The next major achievement of the infant universe was the production of small compound nuclei, such as helium, deuterium, and lithium. Nuclei of these light elements were generated by nuclear fusion reactions during the first few minutes of time. Larger nuclei, including the carbon and oxygen that provide the basis for life, were produced much later in the hot interiors of stars (as discussed in the following chapter). The creation of heavy elements from lighter ones, known as *nucleosynthesis,* greatly alters the matter content of the universe.

Energy is the basic concept driving nucleosynthesis, which is a process of nuclear fusion. Up to a point, larger nuclei have less rest mass energy per particle than the constituent particles contain separately. For example, the rest mass energy of a helium nucleus, made of two protons and two neutrons, is lower than the total rest mass energy of the four particles when they are well separated. This deficit of mass energy associated with the helium nucleus must be accounted for. During the fusion reaction that synthesizes the helium nucleus, the missing mass is turned into energy and released in accordance with Einstein's famous formula $E = mc^2$. Nuclear fusion is the mechanism underlying hydrogen bombs, energy generation in the Sun, and nucleosynthesis in the early universe.

The protons and neutrons in the atomic nucleus are held together by the strong force, which attracts the constituent particles but acts over only a short range. On larger size scales, the electromagnetic force is stronger because it acts over a long range. For example, when a proton interacts with a deuteron (a simple nucleus containing one proton and one neutron), the electromagnetic force is repulsive and acts as a barrier to fusion by pushing the interacting particles apart. In order for the proton and the deuteron to fuse together, they must be close enough so that the strong force overwhelms the electromagnetic force. At sufficiently high temperatures, the particles have enough kinetic energy to achieve the requisite proximity. The temperature cannot be too high, however, otherwise the newly fused nuclei fly apart as soon as they are made. This compromise sets up a range of temperatures over which nuclear fusion can take place.

Relatively early in cosmic history, about one second after the big bang, the background temperature of the universe dropped to ten billion degrees kelvin. With a density 200,000 times the density of water, the universe was cool enough to begin fusing protons and neutrons into light atomic nuclei. A great deal of helium was synthesized, with smaller admixtures of deuterium and lithium. This nuclear activity continued for a rather short span of time, about three minutes. At this juncture, the temperature of the ever-expanding universe fell to one billion degrees kelvin and the density dropped to only 20 times the density of water. Nuclear reactions then abruptly shut down and the phase of nucleosynthesis came to an end.

Although nucleosynthesis produced most of the helium that exists today, the fusion of elements did not proceed to completion during the three-minute window. Most of the universe, about 75 percent by mass, emerged unprocessed in the form of single protons (hydrogen). The rate of nuclear reactions is determined by the temperature and density of the universe. As the universe expands and cools, nuclear reaction rates rapidly decrease and eventually cease altogether. Almost no nuclear reactions take place at low temperatures—notice the conspicuous absence of nuclear fusion at room temperature. Primordial nucleosynthesis was thus a kind of cosmic race. The starting gun went off when the universe was about one second old, when the temperature first became low enough for nuclei to survive. The process of nucleosynthesis and the production of elements began. The race ended approximately three minutes later (slightly less than the time required for an Olympic 1500 meter race) when the expanding universe became too cool to drive nuclear fusion reactions.

If nucleosynthesis in the early universe had continued indefinitely, all of the protons and neutrons would have eventually fused into iron. But why iron and not larger nuclei? Although energy is released by adding together small nuclei to build larger ones, nuclei heavier than iron can release energy through *fission* into smaller daughter nuclei. The fission of uranium works in this manner and provides an energy source for both electric power plants and atomic weapons. Since the fusion of light nuclei releases energy, and the fission of large nuclei also releases energy, an intermediate-sized nucleus must have the lowest possible energy. This most energetically favored nucleus is iron.

As shown in the figure opposite, the theory of nucleosynthesis makes an important prediction. The abundances of the elements produced by the early universe depend on the total amount of ordinary baryonic matter. In order for the predicted quantities of light elements to agree with the values we actually observe, the total abundance of baryons (the protons and neutrons which make up the nuclei) in the universe must fall within a rather narrow range. For the predictions of nucleosynthesis to be consistent, the baryon abundance must lie between 2 percent and 8 percent of the total density required to close the universe. If the total abundance of baryons in the universe were higher than 8 percent of the closure value, the early universe would produce more helium than we observe in the universe today. Similarly, if the abundance of baryons were lower than 2 percent, the amount of helium would be

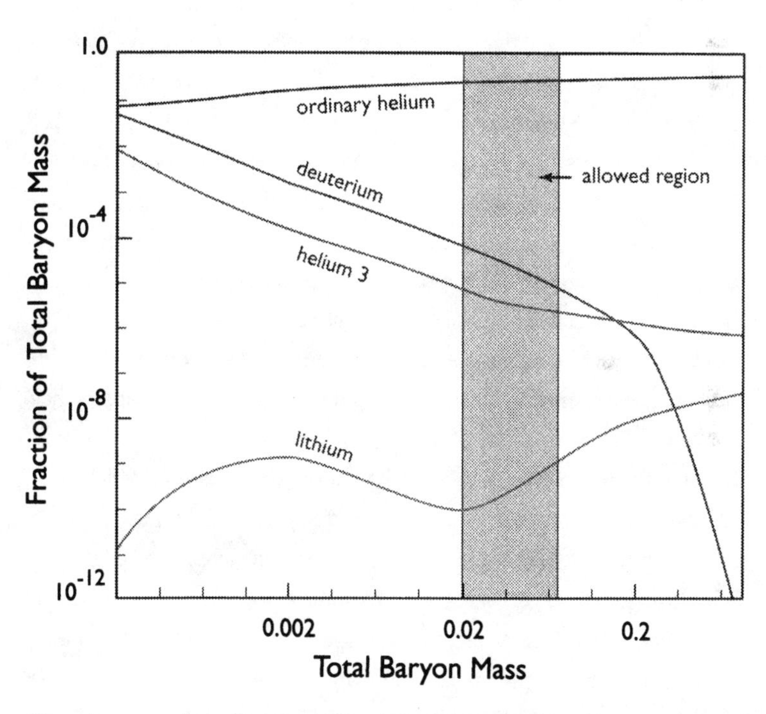

The various curves here show the predicted amounts of the light elements helium, deuterium, and lithium arising from nucleosynthesis during the first few minutes of the universe's existence versus the total abundance of ordinary baryonic matter. The total baryonic mass is written as a fraction of the critical density, so that a value larger than one is required to close the universe. Since the allowed values for the total baryonic mass are much lower than the total mass observed in our universe, some type of non-baryonic matter must also be present.

too low. It is remarkable that a small range of values for the baryon abundance can reproduce the correct amounts of helium, deuterium, and lithium.

DARK MATTER

The rather narrow allowed range for the total abundance of baryons has important consequences for the inventory of the universe. Matter composed of protons and neutrons is baryonic, the stuff of everyday existence: specks of dust, human beings, and the Earth itself. Now if nucleosynthesis tells us that this ordinary matter makes up less than about 8 percent of the total mass in the universe, for a flat universe, what makes up the remaining mass? The answer is *dark matter,* and this exotic substance plays an increasingly important role as our universe matures.

Several independent methods are used for estimating the total amount of mass in the universe. Let's first consider the mass contained in stars. Stars are the most obvious source of material and they are relatively easy to detect due to all the light they emit, even when they live in distant galaxies. Furthermore, stellar properties are fairly well understood. The total amount of stellar mass in the universe appears to be surprisingly small, however, less than 1 percent of the closure density. In light of stellar observations and the theory of nucleosynthesis, the universe must contain two to eight times more mass in baryons than is accounted for in ordinary stars. As a result, some portion of the baryonic matter in the universe must reside in dark forms that emit little radiation. In other words, some *dark baryonic matter* must be present in the universe, in addition to any dark matter made from more exotic, nonbaryonic, material.

Moving up to larger scales, we can estimate how much mass resides in the galaxies. Astronomers measure the speed at which stars are orbiting about the centers of other galaxies. By using these measured speeds and the laws of gravity, we can estimate how much material is present. The larger the observed orbital speed, the more mass required. This accounting procedure implies that most of the mass contained in galaxies resides in the outer galactic halos, and that these galactic halos contain perhaps a hundred times more mass than the stars themselves. The total mass of galactic halos thus accounts for about 10 percent of the closure density. This mass fraction is slightly larger than that allowed in baryonic form by the theory of nucleosynthesis. As a result, we have strong suspicions that some portion of the mass in galactic halos resides in more exotic form, and not in ordinary baryonic material.

Astronomical observations also determine the amount of mass distributed over regions the size of galaxy clusters. In this case, we measure how fast the galaxies themselves are orbiting about the cluster centers. We can also measure how the clusters bend light rays as they pass through. Although somewhat uncertain, these measurements indicate that galaxy clusters contain roughly 30 percent of the mass required for closure. This large reservoir of material is 4 to 15 times more massive than the total amount of baryonic matter, which means that a substantial fraction of the matter in the universe must reside in nonbaryonic form. As a general trend, as ever larger volumes of the universe are weighed, the dark matter makes its presence more strongly felt.

Just what *is* this dark matter anyway? Although a definitive answer has not yet emerged, we have several clues suggesting that dark matter particles must be weakly interacting. In other words, the particles feel only gravity and the weak nuclear force. They are impervious to the strong force and the electromagnetic force. This requirement, coupled with the survival of these particles to the present day, greatly limits the allowed mass range for the dark matter particles.

The masses for the particles fall into two different categories. The first regime includes particles with masses of about 10 to 100 times the mass of a proton. Such heavy particles are relatively slow moving and are generally known as *cold dark matter*. The second possible regime includes lighter particles, with approximately a billion times less mass. These light particles, which generally have relativistic speeds when their abundance is determined, are known as *hot dark matter*. Our universe may contain both types of dark matter particles, but the dark matter particle populations have not been measured. Ongoing experiments hope to shed some light on this dark matter question.

At the present time, weakly interacting dark matter particles affect the universe mostly through their gravitational attraction. Through gravity, dark matter assists in the formation of galaxies and clusters, and helps to guide their present-day motions. Because the time required for dark matter to interact greatly exceeds the current age of the universe, these particles are now largely inert. As we shall see, however, the interactions of dark matter particles become increasingly important as the universe grows ever older. At some time in the distant future, these particles will provide the dominant energy generation mechanism for the entire universe.

2

THE STELLIFEROUS ERA

$6 < \eta < 14$

STARS ARE BORN, EVOLVE TO POWER THE UNIVERSE WITH NUCLEAR REACTIONS, AND THEN DIE BY SPECTACULAR FIREWORKS OR BY SLOWLY FADING AWAY.

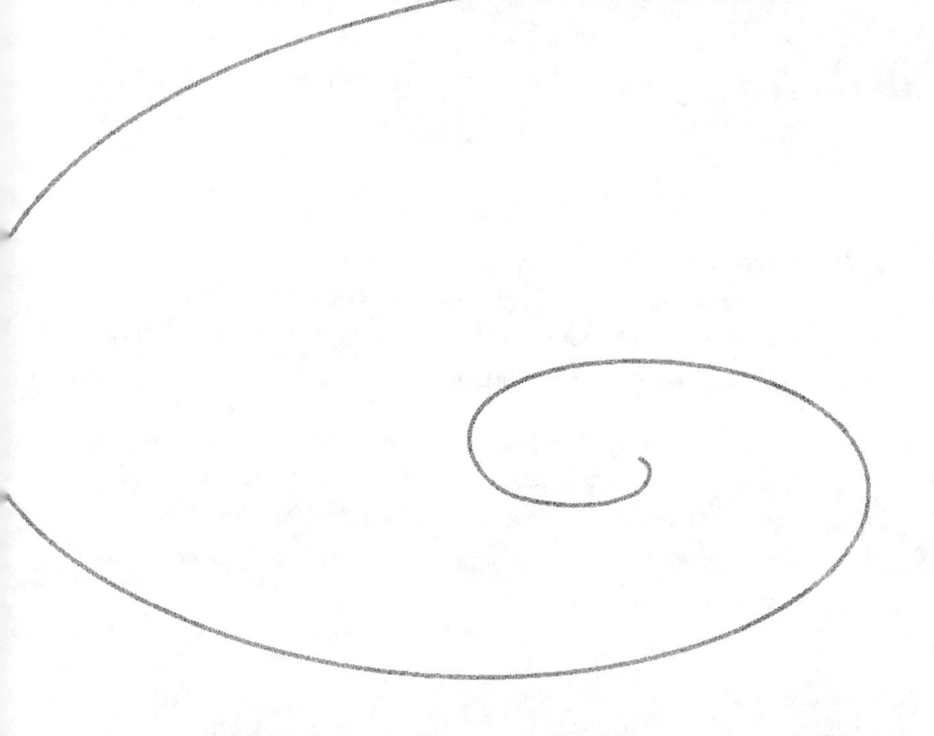

July 11, 1991, El Pescadero, Baja California Sur, Mexico:

The partial eclipse phases leading up to totality lasted for more than an hour. Even as an ever-larger fraction of the Sun became obscured, the change was so gradual that one's eyes could adjust continuously. Any slackening in the daylight remained undetectable until about fifteen minutes before totality, when more than 90 percent of the Sun's face became covered.

Due to the reduced sunshine over a patch of the earth as large as the diameter of the Moon, the morning was unusually cool for a Mexican July. By 10:00 A.M., the temperature was only in the low seventies and then slowly dropped as the eclipse progressed. When the daylight finally began to dim noticeably, the air seemed almost chilly. The surface of the ocean looked dull and flat, but without the slate gray color of a cloudy day. Cumulus clouds billowed over the distant spine of mountains like an accelerated film.

Suddenly, the dunes were awash in strange shadowy ripples, like the bottom of a swimming pool at noon. The ripples drifted slowly across the sand, their contrast flickering. These shadow bands, among the rarest of natural phenomena, were produced by the special combination of a crescent sliver of Sun and unusual turbulence in the upper atmosphere. The bands persisted for less than a minute, and then seemed to evaporate. The wind grew stronger.

With only a minute left, the sky grew darker every second. The air was suddenly alive with flapping fruit bats that had been fooled into emerging by the unnatural dusk. A dangerous stray glance at the Sun gave a moment's impression of a starlike point. With five seconds left, the black shadow of totality swept across the water at more than one thousand miles an hour.

The starlike impression of the Sun was superseded by the disk of the Moon easing into place. A final, fleeting, brilliant burst of light flashed out as the Sun shone through a valley on the limb of the Moon. Totality descended, the stars leapt out, and the nebulous electric blue corona arced away from the black disk.

The intense drama of a total solar eclipse stirs our emotions because human civilization is crucially dependent upon light and heat from the Sun. The Sun has driven the evolution of life on our planet, and it continues to sustain our biosphere. We derive an odd innate comfort from the awareness that nothing we do on Earth can have even the slightest effect on the Sun. It will shine down tomorrow. It will shine for the rest of our lives, and it will shine long after we're gone. The Sun, however, will not shine forever.

The twelve-billion-year lifespan of our Sun is already almost half over. In another six billion years or so, the Sun will exhaust the hydrogen stored in its core, and its long struggle against the force of gravity will enter a new, more desperate, phase of competition. As its supply of hydrogen drops to crisis proportions, the core of the Sun will compress under its own weight, and the surface layers begin a long swell toward the orbit of Venus. In the process, enough radiation will pour out of the solar surface to completely sterilize the Earth.

The fate of the Sun holds obvious ramifications for our own particular long-term future. On a more universal scale, however, the Sun is but one of a billion trillion stars that lie within our cosmological horizon. These luminous stars are now the most important building blocks of the universe. The stars light up the night sky and make up the galaxies. The stars forged the oxygen, silicon, and iron out of which Earth is primarily made. Starlight provides us with most of the information we have concerning the current state of our universe. We thus live in the midst of the *Stelliferous Era*, which will last for another hundred trillion years as normal hydrogen-burning stars continue to shine in a leading role.

GALAXY FORMATION

After the fireworks of the first three minutes, the universe settled into a phase of relative stasis. For the next 300,000 years, space was filled with a nearly featureless sea of hydrogen and helium nuclei, photons, and free electrons, all in a state of constant interaction known as thermal equilibrium. Throughout this peaceful epoch, the universe was expanding and cooling, but the all-pervasive bath of light prevented the growth of structures of any kind. This fiery expanse contained no galaxies, no stars, no planets, and no life forms. The only features breaking the monotony were very slight undulations in the background density. These perturbations were relics from the early universe, most likely the inflationary phase, which now lay in the distant past.

This uncomplicated epoch ended abruptly when the universe cooled down to a temperature of about 3000 degrees kelvin. At this temperature, electrons and atomic nuclei combine to form ordinary atoms, mostly hydrogen. The sea of background radiation loses energy as the universe expands. As the radiation temperature plummets, the photons suddenly lack the energy required to separate electrons from the nuclei, and the particles combine to form neutral atoms. After this combinatory event, the sea of photons has almost no further interactions with matter and streams unhindered through space. The newly complete hydrogen and helium atoms are then free to collapse under the influence of gravity. The ensuing collapse produces vast aggregations of stars, gas, and other matter which we now know as galaxies.

The basic ingredient of the galaxy formation process is conceptually quite simple: Gravity pulls matter together into galaxy-sized structures. The galaxies we see today collapsed from regions which were originally only slightly denser than neighboring regions. Loosely speaking, when a slightly overdense region with the mass of a galaxy begins to collapse, dissipation and cooling halt the collapse as the material approaches a galaxy-sized structure. Most of the initial galactic seed regions were endowed with some slight degree of rotation, that is, with a small amount of *angular momentum.* Since angular momentum is conserved during the subsequent collapse, rotating disklike structures naturally tend to form. These galactic disks support the beautiful spiral patterns that we often associate with galaxies, as shown in the figure on the next page.

Galaxies have populated the universe almost from the start of the Stelliferous Era. With instruments such as the Hubble space telescope, we can actually see galaxies as they appeared when the universe was only a billion years old. This clarity of hindsight is made possible by the finite travel time of light. For example, when we look at the Andromeda galaxy, faintly visible on very dark autumn evenings as a small fuzzy patch in the sky, the light that we see was emitted from stars in Andromeda about two million years ago. This light has been traveling to our eyes for longer than humans have existed as a species. Using larger telescopes to view more distant galaxies, we can literally look back into time at the past history of the universe. Starlight from the farthest known galaxies has been traveling toward us for more than twice the age of Earth!

STAR FORMATION

The first stars were born at about the same time as the first galaxies. At the present epoch, stars form within molecular clouds, vast aggregations of molecular gas residing in galactic disks. These clouds, which often contain the mass of a million Suns, are much denser and colder than the surrounding interstellar gas. Stellar nurseries in nearby molecular clouds, such as the Eagle nebula shown in the first panel of the figure on page 38, provide an environment where we can observe the star formation process in action.

Stars are born out of the collapse of molecular cloud *cores,* small subcondensations which are scattered throughout the much larger volume of a

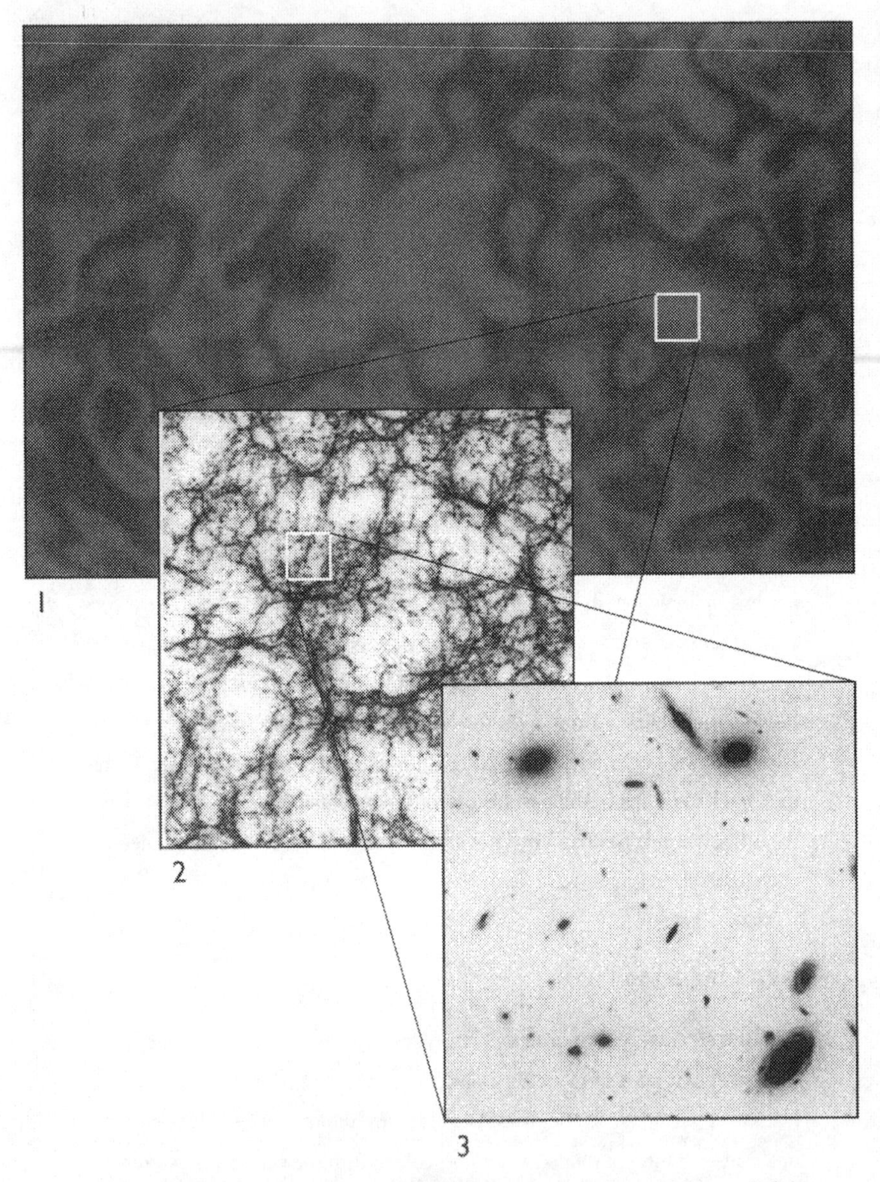

The cosmic fireball stemming from the big bang contained regions which were slightly denser than their surroundings. We can see the signature of these density variations as ripples in the cosmic microwave background radiation (first panel). Later, the denser regions collapse to form sheets and filaments; this process has been simulated on computers (second panel). Eventually, the sheets and filaments fragment and collapse further to make the galaxies and clusters that we see today (third panel).

cloud. These core regions are threaded by magnetic fields, which provide a vital source of pressure to help support the cores against gravitational collapse. However, the cores cannot hold themselves up indefinitely. The magnetic fields gradually diffuse outward and the central regions grow increasingly concentrated. As the magnetic fields abandon the core, it grows too dense and heavy to hold itself up, and the stage is set for a rapid phase of collapse. Soon after the inevitable catastrophe of collapse begins, a small pressure-supported *protostar* appears at the very center of the collapse flow. This stellar seed is destined to grow into a full-fledged star.

The molecular cloud cores that give birth to stars are never completely at rest—these cores spin extremely slowly, about one rotation per million years. This slow rotation imbues the system with a substantial amount of angular momentum. In order to conserve this angular momentum, a molecular cloud core must spin ever faster as it collapses. As a result, not all of the mass falls directly onto the nascent star. A substantial fraction of material collects around the forming star in an accompanying circumstellar disk, roughly the size of our solar system. This nebular disk of gas and dust provides a friendly environment for making planets.

During the main collapse phase, the central protostar and its nebular disk are surrounded by an inward flow of gas and dust. This infalling envelope is thick enough to largely obscure an outside view of the forming star. The original visible radiation emitted by the central star is reprocessed so that forming stars can only be observed at infrared wavelengths, radiation invisible to human eyes. For this reason, bona fide forming stars were not unambiguously identified until the 1980s, when advances in infrared technology catalyzed their discovery.

As a protostar evolves, both its mass and its power output increase. The forming star develops a strong stellar wind which blows out through the torrential rain of gas falling inwards toward the star. When it first breaks out, this outflow is focused into narrow jets, and most of the gas flowing near the star is still directed inward. The outflow jets gradually flare out, however, and start to clear away the sheath of infalling material. In time, the star becomes less deeply embedded in its molecular cloud core. The outflow eventually separates the young solar system from its parental core and the star is newly born.

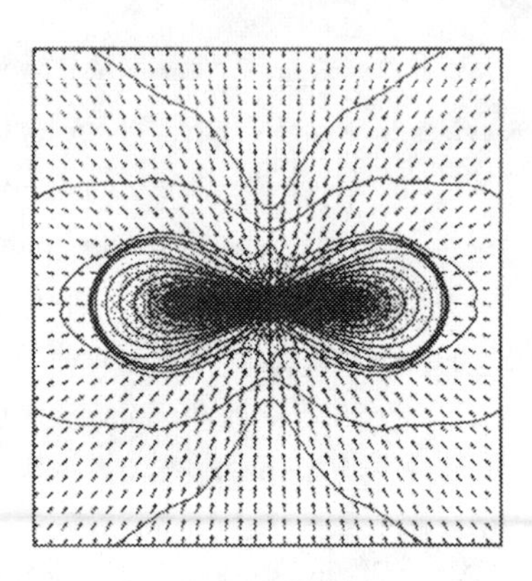

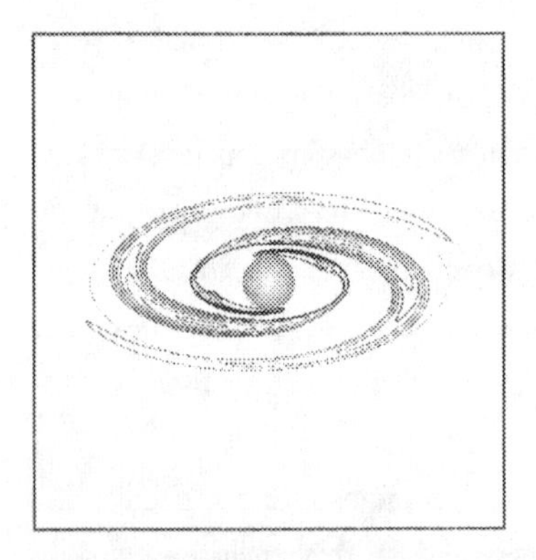

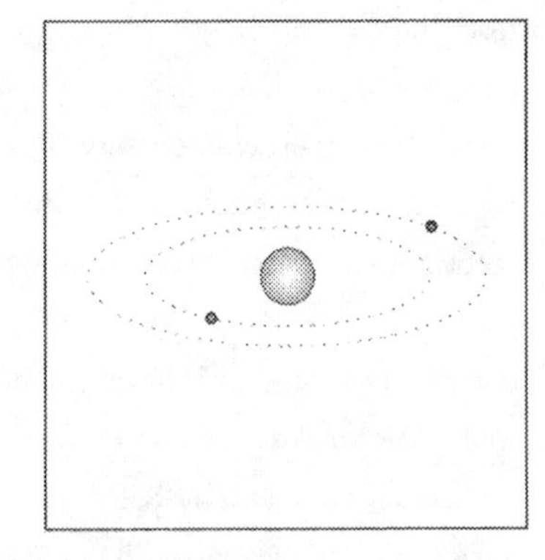

In the first stage of star formation, protostars condense out of the cores of molecular clouds (shown schematically in the first panel). The collapse of a molecular cloud core produces a central protostar surrounded by a disk, shown here in a computer simulation (second panel). The disk starts off with a great deal of mass and generates spiral density waves (third panel), which drain additional gas onto the star. After several hundred thousand years, the remaining disk material can give birth to new planetary systems (shown schematically in the fourth panel).

For the next several million years, this brand new solar system retains its circumstellar disk, where planets slowly accumulate into alien worlds.

Although newly created stars shine very brightly, they do not initially have the proper internal configuration to generate energy through nuclear fusion of hydrogen into helium. At the beginning of their lives, stars derive most of their energy from gravitational contraction. As the star shrinks, its central core heats up, and hydrogen fusion is eventually initiated. Once sustained nuclear fusion reactions finally begin, the star is fully formed.

RIGHT HERE RIGHT NOW

After more than ten billion years of galaxy formation and star formation, we arrive at the present moment. As you read these words, you are sitting somewhere on, or perhaps near, the surface of an 8000-mile-wide planet orbiting an ordinary star. It is useful to pause and take stock of the surroundings.

An airplane flight over our home world provides an interesting perspective. You can see the patch of Earth that provides the focus of your everyday life. Each of us is intensely familiar with the part of our planet that intersects our daily routine: perhaps the grooved pavement on a certain stretch of highway, the bark of a tree in the backyard, or the shadowy concrete canyons gaping between skyscrapers. Flying above all this, you can see how these personal microworlds meld into the overall surface of Earth. Suburbs give way to fields, and highways snake off into the distance. This airborne view hints at the scale of Earth and verifies that we live on the surface of a giant sphere.

Now let's engage in a tremendous mental leap. Imagine that the entire Earth is the size of a grain of sand. A single sand grain is just large enough to see; a large sand grain is barely large enough to feel. This demotion of Earth to the size of sand is akin to a demagnification by a hundred billion. At this scale, the Sun is the size of a dime, and the distance between the Sun and Earth is roughly five feet. Venus and Mercury are additional sand grains lying between Earth and the Sun. Jupiter is the size of a small pea, and lies about 26 feet away. Pluto, the most outlying planet in our solar system, lives two hundred feet in the distance.

Next, make a second, more tractable leap. Imagine shrinking the dime-sized Sun down to the size of a grain of sand. Earth, demagnified by the same factor, is now a microscopic speck, and its orbit traces a circle about an inch in diameter. The distance to Pluto shrinks to two feet. At this scale, the nearest star system—containing Proxima Centauri and Alpha Centauri A and B—is two *miles* away. Stars in the galaxy are like sand grains with miles of space between them. One can hardly overemphasize the relentless emptiness of our present-day galaxy. And the galaxy is a million times denser than the universe as a whole.

Our galaxy contains about one hundred billion stars. If we continue to imagine that each star is the size of a grain of sand, then all of the stars in the galaxy could be packed into an ordinary shoe box. But the stars are not packed close together. To obtain a better feeling for the size of the galaxy, we should scatter the shoe box full of sand across the distance separating Earth and the Moon. Indeed, a photograph of a galaxy can impart a misleading impression. The glowing catherine wheel of stars in such a photograph is the product of a long time exposure taken through a large telescope. In reality, galaxies, even nearby ones like Andromeda, are so dim that they can hardly be seen with the naked eye against the black sky.

With a third mental contraction of scale, we obtain a feeling for the size of the entire visible universe. Imagine that our galactic disk is deflated to the size of a dinner plate. At this scale, the Andromeda galaxy is the size of a second dinner plate and lies suspended several meters away. Galaxies spread out in every direction, and sometimes form clusters. The galaxies and clusters group together to make up filamentary walls that enclose sparsely populated voids, up to a kilometer in size. The entire visible universe extends many kilometers in every direction. At the present time, the visible universe contains roughly the same number of galaxies—ten to one hundred billion—as the number of stars in a single large galaxy. The number of stars in the visible universe is thus similar to the number of sand grains in the dune field shown in the figure opposite.

The visible universe represents the extent of what we can currently observe, but it does not encompass the entire cosmos. Even though regions outside the observable universe are too far away to affect us, such regions nonetheless exist and probably contain similar types of stars and galaxies. As the universe grows older, our cosmological horizon is pushed out, and the visible universe grows larger. Thus, as the future unfolds, more of the cosmos comes into view.

The number of grains of sand in this picture is roughly comparable to the number of stars lying within our entire observable universe in the current cosmological epoch.

INTRODUCING THE CAST OF STELLAR CHARACTERS

The present-day universe is filled with stars and star formation takes place quite readily in the galaxy. But not all spheres of gas are destined for stardom. True stars are confined to the rather narrow mass range from about one-tenth of a solar mass to about one hundred solar masses (where the term solar mass refers to the mass of our Sun).

To sustain nuclear reactions in their interiors, gaseous celestial bodies must contain at least 8 percent of the mass of the Sun. The Stelliferous Era is rife with failed stars, usually called *brown dwarfs,* that are too small to generate nuclear power. Wherever bona fide stars are forming, these brown dwarfs tend to form as well. Their initial collapse from molecular clouds endows them with a dull red warmth, and over the course of billions of years they slowly cool into obscurity. Brown dwarfs effectively store away unprocessed hydrogen fuel. When the Degenerate Era arrives, this investment finally pays off. The brown dwarfs will be hoarding the most significant reserves of hydrogen, and their stock will rise accordingly.

At the other end of the mass range, stars more than a hundred times heavier than the Sun are highly unstable. As soon as an overly massive star is put together, it inevitably destroys itself. The star either generates so much energy that it literally tears itself apart, or else it collapses catastrophically under its own weight and becomes a black hole.

The allowed mass range for stars is much smaller than the conceivable range of masses. Stars live in galaxies which have masses of about one hundred billion (10^{11}) solar masses, and stars are made of hydrogen atoms which have masses of about 10^{-24} grams or about 10^{-57} solar masses. Thus, in principle, galaxies could build objects anywhere in the mass range from 10^{-57} to 10^{11} solar masses, a factor of 10^{68} in mass scale! And yet, stars exist in a range whose mass is allowed to vary only by a factor of a thousand.

The modest range of stellar masses can be contrasted with the huge range in mass sampled by life forms here on Earth. A tuberculosis bacterium has a size of about 1 micron and hence a mass of approximately 10^{-12} grams. Closer to the other end of the spectrum, the blue whale tips the scales with a mass of about a billion grams. The largest plants (such as a contiguous grove

As the long-term future of the stars unfolds, basic stellar characteristics such as mass, luminosity, and temperature play a fundamental role. One useful method for charting the course of stellar evolution is to use the *Hertzsprung-Russell diagram,* a particular type of graph developed independently in the early 20th century by astronomers Henry Norris Russell and Ejnar Hertzsprung. In the Hertzsprung-Russell diagram, the luminosity or power output of a star is plotted on the vertical axis, and the temperature on the horizontal axis. For historical reasons, the temperatures along the horizontal axis are plotted backwards, so that they increase toward the left.

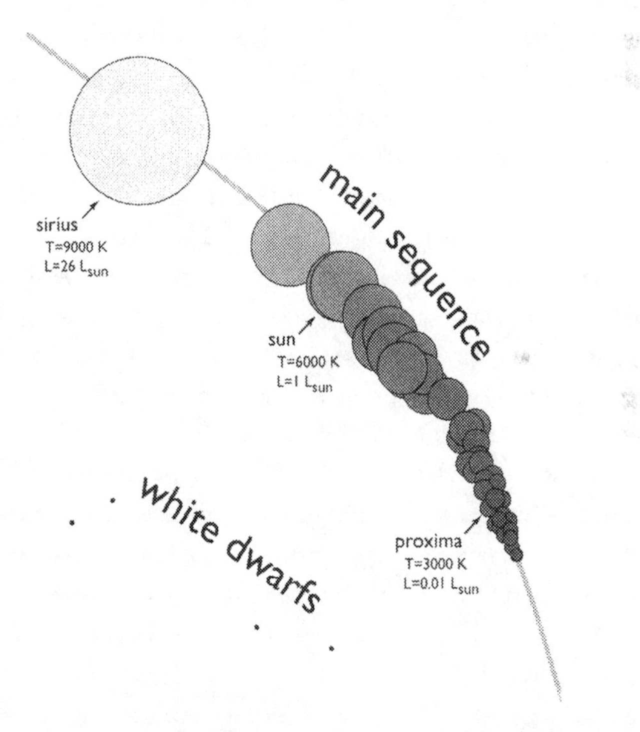

In this Hertzsprung-Russell diagram, the power output of the stars increases toward the top of the diagram, and the surface temperature of the stars increases toward the left of the diagram. The fifty nearest known stars are shown here, with the physical sizes of the stars indicated by the size of the spheres. Notice that most of the Sun's nearest neighbors are small stars (called red dwarfs) and hence the Sun is rather bright compared to most of its neighbors.

(continued)

When we plot various stars on the Hertzsprung-Russell diagram, many stars fall along a well-defined line known as the *main sequence*. This trend is no coincidence. Stars that lie along the main sequence have the proper internal configurations to support the fusion of hydrogen into helium. Stars spend most of their lives in this hydrogen-burning state, and hence most stars lie on the main sequence.

of Aspens) and the largest known fungi (a miles-wide underground specimen residing in Michigan's Upper Peninsula) can reach masses that are much larger still. Individual organisms on Earth span more than 21 orders of magnitude in mass, far vaster in scope than the mass range available to the stars.

During the hydrogen burning phase of evolution, massive stars are a great deal brighter than small stars. A star with ten times the mass of the Sun (such as Spica in the constellation Virgo) is ten thousand times brighter than the Sun, whereas a commonplace small star with one-fifth of the Sun's mass (such as Barnard's star, a nearby red dwarf) is more than one hundred times dimmer. Because massive stars are so much brighter and more energetic than their lower mass counterparts, the massive stars generally dominate the light output of the galaxy. In our night sky, for example, all but one of the fifty brightest stars are more massive than the Sun.

Since the stars visible to the naked eye are almost exclusively more massive than the Sun, one might think that the Sun is a rather small star. This conjecture, however sensible, is simply not true. Of the fifty nearest known stars, the Sun ranks a very respectable fourth and is thus relatively large. The lower-mass stars, mostly red dwarfs with less than half the mass of our Sun, completely dominate the stellar population by number. They also contain most of the total mass in the stellar inventory. Despite their ubiquity, the low mass neighbors of the Sun float through the skies practically unseen. Their meager output of light pales in comparison to the rare and distant massive stars.

We can be more precise about the masses of the stellar population. The distribution that specifies the percentage of stars that are born into each part of the stellar mass range is known as the *initial mass function.* This mass distribution determines the light output of galaxies, their chemical content, and ultimately the inventory of stellar remnants left over at the end of the Stelliferous Era.

All stars fritter away their youth like pyromaniacs, continually burning hydrogen into helium. The lower-mass stars consume their hydrogen fuel in a very frugal manner and survive for a very long time. Massive stars, on the other hand, burn themselves out rapidly and dramatically. Indeed, one might liken a red dwarf to an extremely poor and miserly hermit, who scrapes along from year to year while spending virtually nothing. In contrast, the most massive stars bear an uncomfortable resemblance to rich and profligate heirs, who run shamelessly through a multimillion dollar estate over the course of a single weekend.

THE FATE OF THE SUN AND THE EARTH

From our terrestrial viewpoint, the most pressing question in stellar evolution is the future that lies in store for our Sun. The Sun shines steadily because its central core is hot and dense enough to sustain nuclear reactions that fuse hydrogen into helium (see the figure on the next page). Nuclear fusion is the same process which provides the devastating energy release in a thermonuclear bomb. Because the fusion process transpires deep within the interior of the Sun, and not in the Bikini Atoll, the Sun transforms its hydrogen into helium in a highly controlled manner. Fusion in the Sun is self-regulating in the sense that just enough energy is produced to offset the ever-present inward crush of gravity. If nuclear fusion did not occur, the Sun would contract into a very hot and dense white dwarf over the course of a few million years.

Radioactive dating tells us that the oldest meteorites, and by extension the Sun and the planets, were formed four and a half billion years ago. After the genesis of our solar system, the Sun contracted over the next few million years until its core became hot enough for hydrogen fusion to take place. These nuclear reactions have thus been powering the Sun for the last four and a half billion years. At the center of the Sun, the temperature is sixteen million degrees kelvin. So far, about half of the hydrogen at the center has been burned into helium. Enough hydrogen remains in the extended core to keep the Sun shining

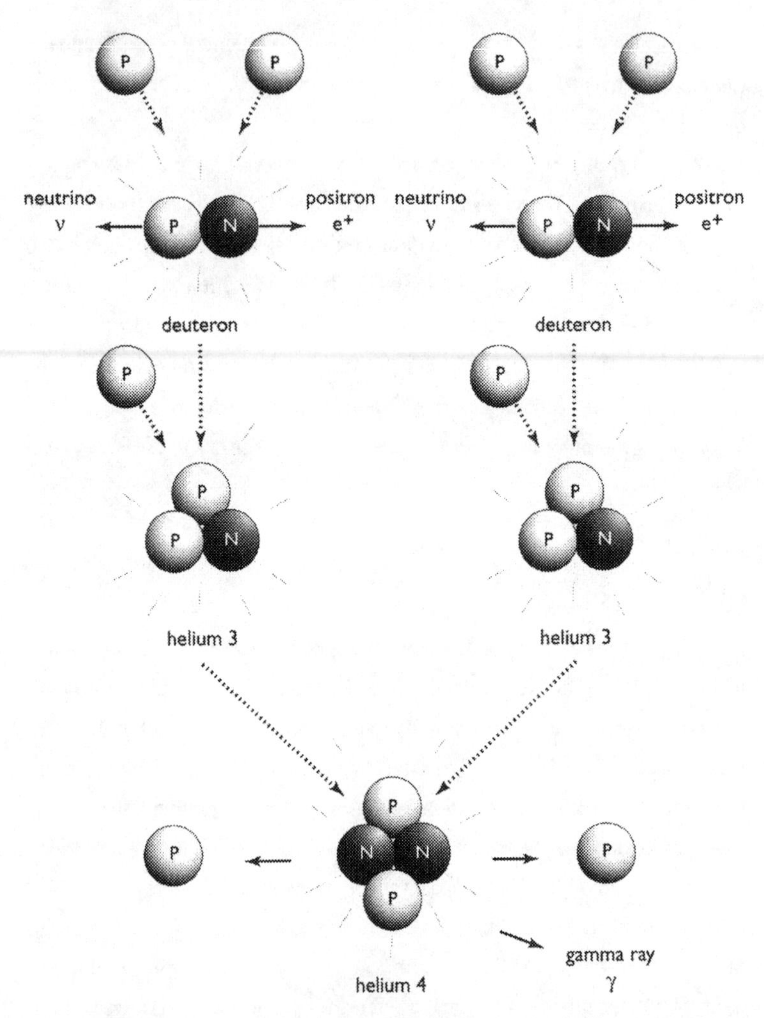

The standard proton-proton fusion sequence shown here is one of the principle nuclear reaction chains by which the Sun generates its fusion energy. As a net result of this process, four protons are converted into one helium nucleus, with extra energy left over to power the Sun.

for another six billion years. Its overall power output must gradually increase as the concentration of helium rises in the center. This increase in luminosity is imperceptible over the course of a human lifespan. Nevertheless, in six billion years, the Sun will be about twice as bright as it is today.

The brighter and hotter Sun presents grave consequences for life on Earth. As the flux of solar energy striking Earth increases, global warming and its unpleasant side effects become increasingly severe. As the oceans grow warmer, their waters hold less dissolved carbon dioxide. Reduced levels of carbon dioxide in the water lead to more carbon dioxide in the air. Although carbon dioxide is only a minor constituent of the atmosphere, this increase in the carbon dioxide concentration has serious implications.

When sunlight reaches Earth, some of the energy is reflected directly back into space. Sunlight that reaches the ground heats up Earth's surface, which then re-radiates infrared light outwards. An atmosphere laced with carbon dioxide and water vapor is largely transparent to visible sunlight, but is almost opaque to the outgoing infrared radiation. The net effect of leaking carbon dioxide into the atmosphere is somewhat akin to putting a lid on a pot of boiling water. Heat that enters the pot has a much harder time leaving.

This so-called *greenhouse effect* can develop into a vicious cycle. Warming oceans release carbon dioxide and water vapor, which lead to further heating, which in turn releases even more gas into the atmosphere. An unstable situation develops. As this runaway greenhouse effect gathers momentum, the oceans evaporate entirely and scald Earth with a sterilizing atmosphere. Although thermophilic bacteria may survive the initial torrid onslaught, all life must ultimately face extinction. A heavy cloud cover might reflect enough sunlight to delay the pending catastrophe, but mass extinction is inevitable. Within a few billion years, our world now green and flowering with life will closely resemble the present-day Venus, with a hellish atmosphere fueled by a runaway greenhouse effect.

On a brighter note, as Earth and its climate slide into an overheated demise, the planet Mars will slowly warm up and become a more hospitable place. In six billion years, the amount of solar energy absorbed by the Martian surface grows to the same level as that received on present-day Earth.

Although a twofold increase in solar brightness is probably sufficient to bring a steamy end to life on Earth, this modest increase in power marks only the beginning of the Sun's fiery old age. After the hydrogen in the solar core has been entirely converted to helium, a shell of material just above the core continues the fusion process. Because the central temperature is too cool to fuse helium into heavier elements, the core lacks an energy source and cannot support the weight of the overlying bulk of the Sun. The ensuing gravitational catastrophe crushes the core until the central pressure can resist the relentless urge toward contraction. As the core shrinks, the temperature of the central region increases accordingly. The heat released by the hotter core is transferred to the overlying layers, where nuclear fusion reactions accelerate. The spent helium core continues to shrink, and the power output of the Sun steadily rises.

Paradoxically, as the Sun becomes more luminous, its surface grows cooler. Today, the solar surface is almost 6000 degrees kelvin. As the Sun swells into a red giant, however, its surface temperature drops to nearly 3000 degrees kelvin, and its color becomes a bright fire engine red. This surface cooling comes about because the excess energy produced near the core is partially deposited in the middle layers of the Sun. These layers are forced to expand and the Sun evolves to become brighter, larger, redder, and cooler than the yellow orb we see today.

As the energy output of the Sun is driven up incessantly, the newly enlarged star develops strong winds. Streams of energetic particles are ejected from the solar surface. These winds resemble the more modest solar wind of today, but they carry away vastly more material. As the Sun ascends into red giant configurations, it will probably lose more than one-fourth of its mass to this powerful wind. As the Sun evaporates, its gravitational pull diminishes, and Earth gradually recedes into a larger orbit, winding up near the present position of Mars. The other remaining planets also find themselves slipping out to orbits of larger radii.

A billion years or so after hydrogen is first depleted in the center of the Sun, the exhausted central core becomes so dense that most of the pressure is provided by degenerate electrons. The term degenerate is used here in the quantum mechanical sense. Electron degeneracy is primarily a consequence of Heisenberg's uncertainty principle. As electrons are forced to occupy smaller volumes, their velocities mount, and the pressure they produce increases. A stellar object supported by this degeneracy pressure is called a white dwarf and is about the radial size of Earth. An expanding red giant basically consists of a central white dwarf surrounded by an extremely deep and diffuse stellar atmosphere. The tiny central white dwarf has a substantial mass, almost half the mass of the Sun. The central core thus reaches an absolutely incredible density, about one million times denser than water.

The severe compression of the core which produces this bizarre dense state is aided and abetted by two curious properties of degenerate material. First, if mass is added to a white dwarf, the additional material causes the white dwarf to shrink in radius. This behavior is completely different from that of ordinary material. If matter is added to an Earthlike planet, which is basically a large rock, the additional material increases the radius of the planet. In contrast, as more mass is piled onto the white dwarf, its radius steadily shrinks.

The second unusual property arises when degenerate matter is heated. An increase in temperature produces neither expansion nor an increase in pressure. This behavior is once again in stark contrast to that of ordinary gases, which increase their pressure and tend to expand as they are heated. Heating a degenerate core to higher temperatures does nothing to alleviate the extraordinarily high densities of the gas.

When the central temperature of the red giant reaches one hundred million degrees, a new sequence of nuclear reactions begins in the degenerate core. Helium nuclei are fused into carbon. The rate of this carbon production depends very sensitively on the temperature. A very slight temperature increase results in an enormous increase in the nuclear reaction rates. This hair-trigger sensitivity, in conjunction with the degenerate core's reluctance to develop more pressure, leads to a runaway situation in which the fusion of helium into carbon escalates out of control. The core of a red giant is rapidly transformed into a colossal helium bomb. For a brief time, the energy production rate of the red giant is comparable to the power produced by all of the stars in the galaxy. This tremendous ensuing burst of energy, known as the *helium flash,* is so powerful that it heaves the dense core out of degeneracy and into a more expanded and stable configuration.

After the helium flash lifts its core out of degeneracy, the Sun enters a relatively stable phase in which helium fuses into carbon in a controlled manner. Because of the energy deposited by the helium flash, the core region is no longer compressed and degenerate. In this new state, the Sun cannot maintain its distended red giant configuration. At the peak of its red giant phase, the Sun will be two thousand times brighter than it is now. Once helium has ignited, however, the Sun becomes smaller, hotter, and less luminous. The relatively quiescent helium burning period lasts for about a hundred million years. During this time, the Sun shines forty to fifty times brighter than it does today, and its surface is a thousand degrees cooler. The temperature on Earth cools back down to several hundred degrees Celsius, and the planetary surface can resolidify. Earth's new crust reveals little hint of the geology, biology, and civilizations that once graced its surface.

LOST IN SPACE

As the Sun churns through its life cycle, Earth's biosphere is scheduled to be completely destroyed by a runaway greenhouse effect in about two billion years. Although this future catastrophe is not an immediate concern, it nonetheless offers a depressing prospect. Can life on Earth somehow outlast the Sun's inevitable luminosity increase?

In a well-known poem by Robert Frost, some believe the world will end in fire, while others say ice. In this astronomical setting, Earth has a slim chance of escaping the fiery wrath of the red giant Sun by becoming dislodged from its orbit and thrown out into the icy desolation of deep space. Other stars living within our galaxy regularly pass by our Sun, and they could possibly blunder into the inner portion of our solar system. In this unlikely event, the disruptive gravitational effects of the close encounter could easily force Earth to abandon its orbit and bolt from the solar system. In this manner, our world may avoid a scalding demise, but would face a frozen future.

Within the next two billion years, the chance of another star passing close enough to throw Earth out of orbit is about one part in one hundred thousand. Although the galaxy contains a huge number of stars, they are spread out through space so thinly that meaningful stellar encounters are extremely rare. Recall that if the Sun were shrunk to the size of a sand grain, the closest star would be a few miles away. At this scale, the nearby stars would be moving at a glacial pace of a foot or two per year relative to the Sun. In a galaxy crawling with such vastly spaced stars, random close encounters are exceedingly rare.

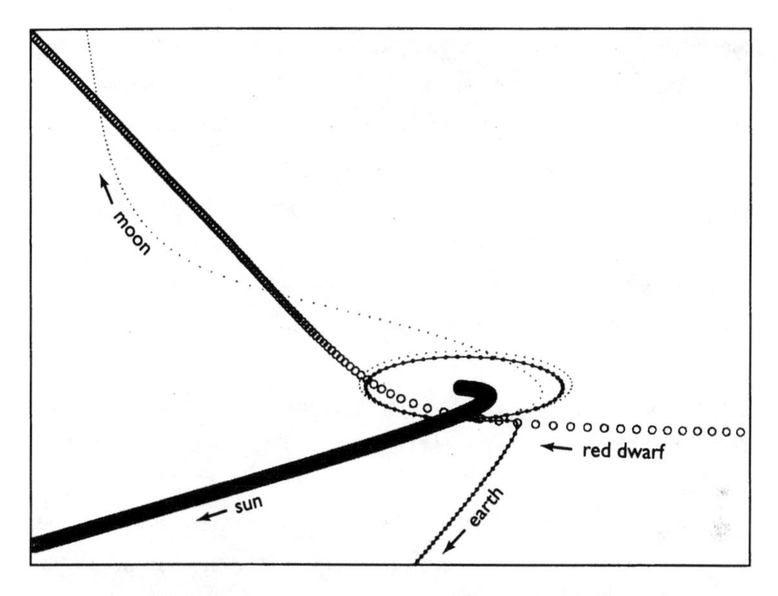

This computer simulation shows Earth, the Moon, and the Sun reacting to a visit from a red dwarf with one-quarter of a solar mass. The simulation predicts that Earth would be ejected from the solar system at high speed. However, the chances of such an event taking place before the Sun turns into a red giant are slim: only about one part in one hundred thousand (1 in 10^5).

One chance in a hundred thousand is rather small, yet people regularly line up to buy lottery tickets that sport far lower odds. Let's examine what might happen if a passing star unceremoniously strips Earth away from the Sun. The figure above shows a typical sequence of events during the close passage of a red dwarf through the solar system. In this particular scenario, Earth warms up slightly as it passes near the interloping red dwarf star. The dimmest red dwarfs produce so little power that Earth is not adversely heated by its brief close shave. The tides from the encounter are horrendous, however, and generate uncommonly huge ocean waves. Except for a few weeks of epic surfing, the main effect of the passing red dwarf is that of a gravitational slingshot. In this case, Earth suddenly finds itself launched away from the Sun at high velocity.

With this speed, Earth crosses the orbit of Pluto within a few years and leaves the solar system. The surface temperature of the planet drops slowly but steadily as the Sun grows ever smaller in the sky. Frost covers the inland jungles of the Amazon and tropical plants freeze in the gathering final twilight. After a year, the Sun appears in the sky as a brilliant starlike point, seemingly devoid of any warmth. The oceans slowly divest their store of heat and freeze into steadily thickening packs of ice. The temperature at the surface drops inexorably. Soon, the entire planet is plunged into a permanent deep freeze, colder than an Antarctic winter. At 77 degrees kelvin, the nitrogen that makes up most of the atmosphere condenses and falls as rain onto the frigid snowy surface. The liquid nitrogen runs downhill along riverbeds, collects on frozen inland basins, and eventually pools onto the frozen ocean surface in a layer several feet deep. The oxygen from the atmosphere also rains out of the frozen skies, and leaves a clear view of distant stars from cold dead cities.

Deep inside Earth, however, these changes are scarcely noticed. The deepest parts of the oceans remain unfrozen in places. Along the mid-oceanic ridge, an undersea range of mountains which circles the globe, the strange web of life-forms living in hydrothermal vent ecosystems continues to function unperturbed. These communities are fed by the volcanic interior heat of Earth and are entirely unreliant on even the slightest ration of sunlight. Impervious to the loss of the Sun, these vent communities continue to thrive until the geological activity of Earth comes to an end, although their character changes substantially as the oxygen supplies dwindle. Geological activity is driven by heat from the decay of radioactive elements (such as uranium) and lasts for billions of years, longer than the remaining lifetime of the Sun. In an interesting twist of irony, life on Earth can actually continue longer if our planet is swept out of the solar system.

For those with a biophilic inclination, the prospect of anaerobic bacteria, and perhaps blood-red tube worms huddled around geothermal vents, is certainly superior to the outright heat pasteurization of the entire planet. Yet the consolation provided by this frigid outcome—whatever the odds—is still not particularly satisfying. The real lottery payoff has longer odds, but is nonetheless possible. Rather than being ejected into deep space, Earth could be passed off into a habitable orbit around an interloping red dwarf. In this fortunate circumstance, life here on Earth could conceivably survive and evolve for trillions of years.

Such a fortuitous exchange is extremely unlikely to occur during the passage of a single star near Earth. The odds are greatly improved if our solar system suffers a close encounter with an impinging binary or triple star. Such an encounter is really no less likely than the close passage of a single star, since more than half of all star systems are either binary or triple.

The close passage of three stars is an extremely complicated affair, and the outcome depends rather delicately upon the initial velocities and locations of the stars. The following figure shows what might happen. In this particular simulation, Earth winds up in an elliptical orbit about a red dwarf. The chances that Earth will be saved in this manner appear to be a bleak one part in three million.

THE FATE OF MASSIVE STARS

The nature of stellar death depends on stellar mass. Single stars that contain more than half the mass of the Sun, but less than eight times the mass of the Sun, are destined to share largely the same fate. At the end of their lives, they disgorge huge quantities of hot gases (called planetary nebulae) and leave behind white dwarfs composed primarily of carbon and oxygen. Only stars heavier than eight solar masses have central temperatures high enough so that nuclear fusion reactions proceed significantly beyond these elements. These more massive stars have a more dramatic end in store.

To fix ideas, let's consider what happens to a 15-solar-mass star. The stellar lifetime is very short; the star burns its central stores of hydrogen into helium within only ten million years. The star then easily ignites helium before the electrons in the central region become degenerate, and just before the helium has been exhausted, the stellar core is composed mainly of carbon and oxygen. The temperature in this core is more than a hundred million degrees, and the density is about a thousand times that of water (1000 grams per cubic centimeter). When the helium at the center becomes entirely depleted, the core begins to contract under its own weight. The density is driven upwards to 100,000 grams per cubic centimeter and the temperature approaches one billion degrees.

When the core of a massive star is heated above a billion degrees, the star

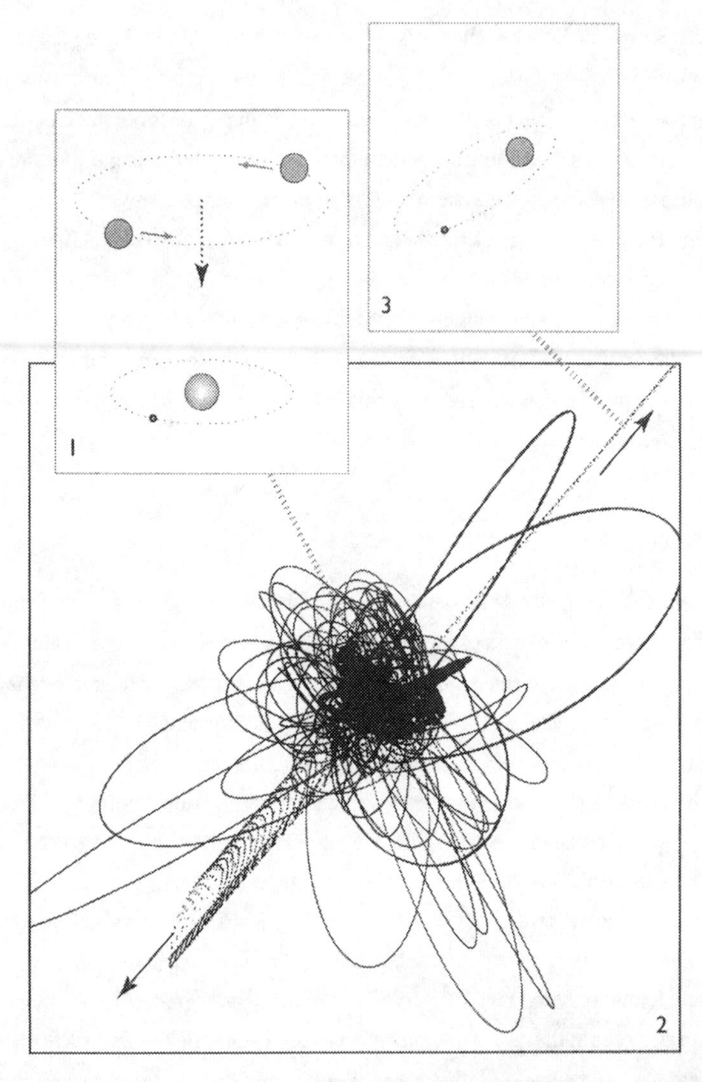

This computer simulation shows the outcome of a close encounter between a red dwarf binary pair and our solar system. As the red dwarf pair drops toward the Sun, Earth is almost immediately handed off to the smaller star and stays with that star for three long, looping excursions. After slightly more than 1000 years, Earth is palmed back off onto the Sun, where it remains for the next 6500 years and suffers many complicated close encounters with the other stars. After 7500 years, Earth is captured into an orbit around the larger dwarf, and soon thereafter, this star escapes. Earth is pulled along in an elliptical orbit that might possibly be habitable. A capture of this sort has about one chance in three million of occurring before the Sun turns into a red giant.

faces a new challenge in its continuing struggle against gravitational collapse. At such high temperatures, prodigious quantities of neutrinos are produced. Because neutrinos have so little chance of interacting with the star as they make their exit, they leak energy into space but provide no pressure support to hold up the star. The net result of these neutrino losses is that not all of the energy derived from nuclear fusion can help support the star against gravitational collapse. Neutrino losses thus hasten the end for massive stars.

When the center of a massive star is hot enough to initiate the nuclear burning of carbon, the contraction and heating of the core is brought to a temporary halt. Carbon burning occurs when two carbon nuclei join together to create an excited magnesium nucleus. The magnesium nucleus can decay in a number of different ways, resulting in the production of neon, oxygen, sodium, and magnesium. This wide variety of fusion products is typical of the complicated later stages of nuclear burning.

One reason why stars spend most of their lives in the hydrogen-burning phase is because the fusion of hydrogen into helium is the most *exothermic* nuclear reaction. Much more energy is released per gram of material when hydrogen is fused into helium than when, say, carbon is fused into magnesium. The diminishing return on successive cycles of fusion into heavier elements, combined with the vastly higher temperature and energy requirements, ensures that the late phases of nuclear burning do not last very long. Within a 15-solar-mass star, carbon burning lasts for only several thousand years. After all the carbon at the center of the star is gone, the core must contract and heat up once again.

After carbon burning has run its course, the structure of a massive evolving star bears a vague resemblance to an onion. A series of distinct layers delineate regions of different chemical composition, from the neon-oxygen-magnesium core of the star out to the surface layers containing unprocessed hydrogen. At the base of each layer, a burning front ignites the nuclear ash produced by the nuclear fusion reactions taking place within the layer directly above (see the diagram on the next page).

In the center of an evolving massive star, a complex maze of nuclear reactions rapidly converts the oxygen and neon into silicon, sulfur, and even larger nuclei. These nuclear reactions continue to build heavier elements until signif-

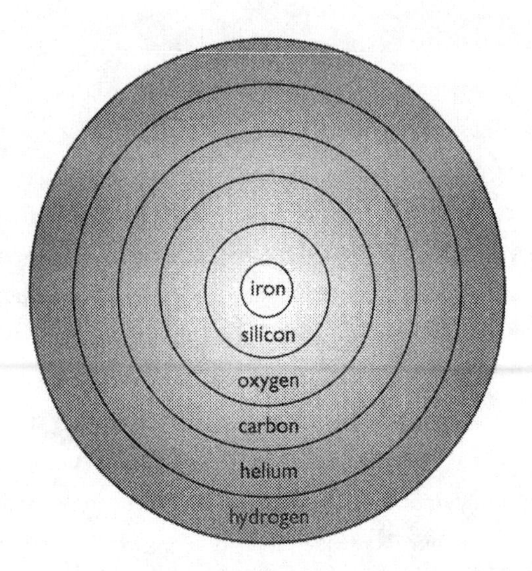

This diagram shows the schematic structure of a very evolved star of high mass, just a few minutes before a supernova explosion. Each "onion skin" represents a layer of nuclear burning near the central inert iron core. Only the central region of the star is shown here; the rest of the star consists of a thick envelope of unburned hydrogen.

icant amounts of iron are present. When the stellar core becomes dominated by iron, yet another problem arises. The fusion of iron into even heavier elements (for instance silver or gold) does not lead to the release of energy, but rather requires the absorption of energy. A star that develops an iron core is unable to squeeze any further energy from nuclear fusion. With a density of about 10 billion grams per cubic centimeter and a temperature of more than a billion degrees, the moribund iron core is unable to support itself against gravity and begins to collapse.

The collapse takes place very rapidly. In a single second, the innermost regions are compressed to colossal densities, approaching 10^{14} grams per cubic centimeter. If Earth were compressed to this density, the entire planet would be about a quarter mile in diameter! Just after the collapse begins, the temperature becomes so high that the background of thermal radiation destroys the iron nuclei in the core. The hard-won iron nuclei are first broken up into helium nuclei, also known as α particles, and finally into protons and neutrons.

This photodissociation process robs the core of its thermal energy, which might otherwise have prevented the collapse. Adding to the star's woes, the thermal photons have enough energy to interact with each other to produce electron-positron pairs. Recall that the positron is antimatter partner of the electron. Although these pairs generally annihilate to produce more photons, they occasionally produce neutrino-antineutrino pairs, which fly outward and remove even more energy from the core.

When the core density approaches 10^{14} grams per cubic centimeter, free electrons and protons combine to produce neutrons and neutrinos. This thick soup of neutrons resembles a single gigantic nucleus. As the collapse continues, this giant ball of neutrons generally reaches a state of maximum density, and then bounces back. This bounce drives an extraordinarily powerful shock wave out through the overlying star.

The collapse of an iron core in a massive star, followed by the bounce at nuclear densities and the ensuing shock wave, is called a *supernova*. In many cases, the shock wave is strong enough to completely blow apart the outer layers of the star. Heavy elements (including gold, lead, and uranium) are synthesized within the shock itself, and the various elements produced by earlier stages of nuclear burning are thrown back out into the interstellar medium.

At the center of the supernova explosion, the dense core of neutrons can remain behind as a neutron star. Alternatively, if the core is heavier than a few times the mass of the Sun, it may collapse into a black hole. The formation of a black hole represents a decisive victory for gravity in its ongoing struggle with thermodynamics and entropy production. But a third possibility also exists. If a sufficiently massive star explodes violently and is efficient at expelling stellar material, then no stellar remnant of any kind is left behind. This possibility represents a clear-cut victory for entropy.

Supernovae are the most dramatic phenomena in stellar evolution. For a brief moment, as the iron core collapses, the mind-bending temperatures and densities at the stellar center revisit conditions which prevailed during the first instants of the primordial universe. The energy release from a supernova is correspondingly spectacular. For a single second, the amount of energy produced by a supernova rivals the total energy emitted by all of the stars within the visible universe. For days after a supernova is touched off, the residual explosion shines as brightly as the entire galaxy that the dying star called home.

THE FATE OF LOW-MASS STARS

Working at South Africa's Union Observatory in Johannesburg in 1916, an astronomer reported the discovery of a faint star in the southern constellation Centaurus. This otherwise unremarkable star, far too faint to be seen with the naked eye, attracted his attention because it was slowly changing its position with respect to other stars in the same part of the sky. This movement indicated that the star might be a close neighbor of the Sun, and in 1917, this suggestion was verified. The distance to the star was measured to be only 4.3 light-years, closer to the Sun than any other known star. Its extremely faint appearance, in spite of its close proximity, made it the intrinsically least luminous star known to astronomy at that time. Proxima Centauri, as the star was later named, is now known to be just one out of billions of red dwarfs which inhabit our galaxy.

These red dwarfs are by far the most common type of star and differ from the Sun in several ways. Proxima has about 15 percent of the Sun's mass, an average density several times that of lead, and a power rating four hundred times dimmer than our Sun. Even this rather modest energy output has a difficult time escaping from the dense stellar interior. The center of Proxima is so opaque that radiation cannot efficiently transport all of the energy produced by fusion out to the stellar surface. To carry its meager luminosity to the surface, Proxima must resort to convection, a process in which the turbulent motion of stellar gas physically carries energy away from the center. Convection can be observed in a pot of water being heated on a stove. Hot water wells up near the center of the pot, divests some of its heat, and then dives back down. This roiling, overturning motion of the water is closely akin to the convection motions that transport energy in low-mass stars.

Inside Proxima's stellar surface, almost the entire star participates in convection and hence the stellar material is continuously mixed together. For example, a helium nucleus forged in the nuclear burning core can expect to visit the surface regions within a relatively short time. This freedom of movement is in direct contrast to our Sun, whose core is *radiative* rather than convective. Helium forming in the center of the Sun never strays far from its place of origin. The Sun's core thus becomes slowly enriched in helium, while the original composition of the outer regions remains unaffected. A low mass star like Proxima is completely convective and maintains access to its entire initial reserve of hydrogen fuel. Complete convection, coupled with an underwhelming power output, allows red dwarfs to survive almost unaltered, long after the higher mass stars have turned into white dwarfs or self-destructed in supernova explosions.

Because red dwarfs have access to nearly all of their hydrogen, they live for an extraordinarily long time. The smallest stars, with about one-tenth of a solar mass, are a thousand times dimmer than the Sun. The nuclear luminosity of a star is ultimately derived by converting some of the stellar mass directly into energy according to Einstein's famous formula, $E = mc^2$. Each helium nucleus is slightly less massive than the four hydrogen nuclei from which it formed. The exact *mass deficit* is seven-tenths of 1 percent; four hydrogens weigh 1.007 times as much as one helium nucleus. Fusing one gram of hydrogen in one second releases 630 billion watts of power, enough to keep a 300-horsepower car running for a month. The hydrogen stored within a newly born red dwarf, with one-tenth of a solar mass, can keep it shining steadily for fourteen trillion years, about one thousand times longer than the current age of the universe. As a star generates energy, it loses a corresponding amount of mass. For a red dwarf, this mass loss is analogous to a fully loaded freight train continually hauling material away from its surface at 100 miles per hour.

Thus far in the history of the universe, red dwarfs have not lived long enough to evolve away from the earliest hydrogen-burning phases of evolution. For this reason, aside from general estimates of their lifetimes, little attention has been given to their ultimate demise. Nevertheless, the fate of the galaxy lies in the hands of the red dwarfs. After the more massive stars have squandered their nuclear fuel and died in a youthful grave, the red dwarfs will continue to shine. These small stars will wheel through space for trillions of years, continuously convecting, slowly shrinking, and gradually brightening. Red dwarfs thus play a major role in the long-term development of the galaxy.

Let's consider the long-term evolution of the lowest-mass star, with only 8 percent of the Sun's mass. As its initial supply of hydrogen is slowly depleted, the star heats up and contracts. The stellar luminosity increases by a factor of 10 and the surface temperature more than doubles. After eleven trillion years, when the star has burned 90 percent of its initial hydrogen supply, convection finally ceases within the central region. Stellar evolution then accelerates and the star quickly burns up the remaining hydrogen stored within its core. Although it develops an exhausted core of helium and an overlying shell of nuclear burning, the star is not powerful enough to turn into a red giant. Instead, as the star grows hotter and smaller, the red dwarf transforms itself into a blue dwarf. After about 98 percent of the initial hydrogen fuel has been consumed, nuclear burning stops, and the star cools down to become an almost pure helium white dwarf. Stars with masses less than 20 percent as large as the Sun follow this same evolutionary story.

A slightly larger star, with about one-quarter of a solar mass, loses its convective core sooner. The conditions inside the star are sufficient to briefly drive the star to a lower surface temperature as the luminosity increases, and the star evolves toward a red giant configuration. Stars born with a quarter of a solar mass are the smallest stars that become red giants.

As the more massive stars die without being fully replaced, the vast aggregate of aging red dwarfs will produce a larger fraction of the galaxy's total light output. Their slow-but-steady luminosity increase will keep the galaxy shining with the light of a billion Suns for trillions of years. After a 0.2-solar-mass star has lived for a trillion years, for example, it will exhibit both the same luminosity and the same surface temperature as our Sun. If this evolved star could somehow change places with our Sun at the center of the solar system, Earth and the other planets would suddenly find themselves in unbound hyperbolic orbits (due to the difference in mass), but the brightness and color of the 0.2-solar-mass star in the sky would be the same as that of the Sun.

Many low-mass stars experience an extended time period during which they shrink and grow hotter, but their total luminosity remains roughly constant. For a 0.16-solar-mass star, this phase occurs after the core has been fused to pure helium, and a hydrogen-burning front is working its way toward the surface. During this period, the star shines with a relatively constant luminosity, about one-third as bright as the present-day Sun, for more than five billion years. This warm phase of steady luminosity allows ample time for life to evolve on any appropriately situated planets. Recall that here on Earth, simple unicellular organisms have evolved into people in less than four billion years. This epoch represents the maximum brightness attained by the star. Before this late blooming period of warmth, any accompanying planets would have languished in cold storage as the star churned through trillions of years of convective evolution.

As the galaxy ages and stellar generations go by, the concentration of heavy elements in the interstellar medium steadily increases. As a result, stars of the far future will contain more heavy nuclei than the stars of today. This forthcoming increase in impurities lowers the minimum mass required for a star to burn hydrogen. When the impurity level reaches several times the current solar value, stellar objects with only 4 percent of a solar mass may sustain hydrogen fusion in their central cores, while thick ice clouds condense in their atmospheres. These bizarre frozen stars can display effective temperatures near the freezing point of water, zero degrees Celsius or 273 degrees kelvin, far cooler than the smallest and coldest stars of today. As these frugal objects slowly burn their hydrogen fuel, with a power output one million times less than the Sun, they achieve commensurate increases in longevity.

THE QUEST FOR EXTRATERRESTRIAL LIFE

A planet orbiting a relatively massive star—such as Deneb in the constellation Cygnus—is not likely to harbor an alien civilization. A 10-solar-mass star like Deneb lives for only ten million years before destroying itself in a supernova. If an Earthlike planet is orbiting Deneb, the planetary surface remains molten, or barely solid, and is subject to intense ionizing radiation from the seething stellar surface. This prospective planet belongs to an extrasolar system very much in the throes of formation. An intense bombardment of planetesimals, meteorites, and comets actively adds material to the planet and continually changes its climate. In such a system, not enough time has elapsed for complex life of any kind, much less an intelligent civilization, to develop.

Two lines of evidence suggest that a long time is required for advanced civilizations to arise. Although Earth has existed for 4.6 billion years, essentially all of this time elapsed prior to the evolutionary ascent of humans. The first technological civilization on the planet has existed for an even shorter period of time, only a few hundred years. Thus, in the one example that we know, the evolution of intelligence required billions of years. The second clue is that we have no indications that any other civilizations exist. In particular, no extraterrestrial societies have contacted us. If technological civilizations could easily arise in short periods of time, then we might expect some relatively nearby star to broadcast detectable signals. The stony silence argues in favor of long time requirements for technological development.

What combination of incremental advances leads to a civilization, and how much time is reasonably required for each step? To start the process, primitive life must emerge. By primitive life, we mean the simplest structures capable of both reproduction and natural selection. According to this definition, a virus would constitute one type of primitive life form on Earth. Interestingly, viruses seem to have appeared later than the first cells. In any case, primitive life has been present on Earth from a very early date. The oldest known sedimentary rocks containing fossils show that life was thriving near the South African coast nearly four billion years ago. The emergence of primitive life on Earth took no more than a few hundred million years, an alarmingly short time period, only a small percentage of the current age of Earth.

If all of the stars were heavier than three solar masses, they would live for only a half-billion years, and life would seldom progress much beyond these primitive unicellular stages. In all likelihood, a galaxy full of three-solar-mass stars would rarely reach the second major milestone on the path to a technological civilization, namely the development of highly complex *eukaryotic* cells.

For the next step, complexity must emerge. Life on Earth required more than three billion years to evolve the fantastically complex molecular machinery which operates in present-day eukaryotic cells, including the cells in the human body. During most of geological time, life remained primarily in the one-celled stage, while becoming gradually more complex on the molecular level. A one-celled amoeba, for example, is enormously more sophisticated than a one-celled E. coli bacterium.

If the smallest stars in the galaxy were 25 percent heavier than the Sun, the maximum stellar lifetime would be about three billion years. In this case, only very rare planets could develop life forms beyond the complex one-celled stage. For example, suppose that the nearby star Procyon, weighing in at 1.4 solar masses, has a twin of Earth. Life may have arisen in the distant oceans of this hypothetical planet, provided that it survived the red giant phases of Procyon's white dwarf companion. Even as we contemplate this issue, complex molecular machinery could well be evolving, in an analogous fashion to what took place on Earth two billion years ago. Unfortunately, this miraculous evolutionary development must be prematurely arrested. Procyon's days as a normal hydrogen-burning star are numbered. In only a few hundred million years, the star is destined to swell into a red giant and efficiently sterilize any currently habitable planets in its solar system.

The third major watershed event leading to the development of intelligent beings on Earth was the emergence of multicellular organisms. Large life forms coordinating cells of specialized function first appeared about 800 million years ago, when Earth was nearly 3.8 billion years old. The *Ediacara fauna,* named for the Ediacara hills in Australia where the best-known fossils from this era were found, seem to be only distantly related to the plants and animals of today. These Ediacara organisms, often resembling pillows or air mattresses, apparently floated on the ocean surface, or inhabited deeper waters, and led the sedentary life of filter feeders. The Ediacara fauna may represent an alternate, and ultimately unsuccessful, evolutionary solution to the problem of coordinating large numbers of cells within a single organism. These exotica may have been decimated by the first predatory wormlike ancestors of today's animal phyla.

Complex animals trace their lineage quite clearly back to the *Cambrian explosion* of 540 million years ago. During a span of only 10 to 20 million years, an intense burst of speciation spawned the earliest known members of almost all of the animal phyla now represented on our planet. The combination of events that sparked the Cambrian explosion remains a mystery and many key questions have not been answered. Most importantly, we need to know whether or not the evolutionary time of four billion years is really necessary for largely single-celled organisms to reach a point where the radiation of complex forms is possible. Was this much time required, or did the Cambrian explosion simply represent the outcome of a random set of triggering events which could have occurred much earlier in Earth's history?

Compared to the age of Earth, 4.6 billion years, evolution from the Cambrian explosion up to the appearance of our technological society has taken place quite rapidly. Was this accomplishment blind luck, or was the establishment of intelligence virtually guaranteed once complex multicellular life was in place? We don't know. But the answer is crucial, particularly for ongoing searches for life beyond our solar system. In most considerations of extraterrestrial intelligence, the starting point is to estimate the number of intelligent civilizations now living in the galaxy. Estimates of this sort attempt to tie together many aspects of astronomy, biology, and anthropology.

To estimate the number of stars in the galaxy that are suitable for life, we can adopt Earth's example and require the stars to live at least 4.5 billion years in order to evolve intelligent creatures. Since the stars must live relatively long, they must be relatively small. By this conservative measure, the largest star that can survive long enough contains 1.15 solar masses. In addition, the stars must not have binary companions capable of disrupting habitable orbits. These requirements are not particularly restrictive. Our galaxy is the home of about ten billion suitable stars, and the universe contains nearly ten billion trillion (10^{22}) within its present horizon volume.

Next, we must determine the fraction of suitable stars that actually have planetary systems. Until a few years ago, this fraction was almost entirely unknown, although it was generally thought to be substantial. In the last few years, a number of planetary systems beyond our solar system have been discovered. The rapid discovery rate argues that planetary systems are a natural and common outcome of the star formation process. As more extrasolar planets are discovered, the fraction of stars with planets is being driven towards unity.

Only some fraction of suitable stars with planetary systems have an Earthlike member in an orbit capable of sustaining life. Although uncertainty creeps into this part of the discussion, our present theories of planetary formation suggest that this fraction is fairly large. Earthlike planets are not difficult to form. Our own solar system contains four terrestrial planets. Vaguely similar terrestrial planets have been discovered in orbit around neutron stars. Each of the giant planets in our solar system—Jupiter, Saturn, Uranus, and Neptune—is escorted by an entourage of rocky moons which presumably formed through a similar process as the terrestrial planets. Given that small rocky planets are easily produced, simple statistics argue that a healthy fraction must live in habitable orbits. Using a conservative assumption for the range of habitable radii, we estimate that 1 percent of suitable stars with planetary systems have Earthlike planets in habitable orbits.

For the fraction of Earthlike planets that actually develop life, the available estimates vary widely. One outspoken astronomer, Sir Fred Hoyle, has proposed that the odds are as poor as one part in $10^{40,000}$. On the other end of the spectrum, more optimistic pundits have boldly given one-to-one odds. Amidst such widely varying speculation, we desperately need real data. The most dramatic evidence with the potential for bearing upon this issue is the recent analysis of a Martian meteorite, which might harbor primitive Martian life. After this piece of Mars was blasted away from the red planet by a meteoritic impact, it flew through interplanetary space for many years and eventually fell to Earth. No consensus has been reached as to whether the meteorite is truly indicative of life on Mars. Each piece of evidence for Martian life in the meteorite also has a possible nonbiological explanation. Furthermore, life could have emerged just once within our solar system, either on Mars or Earth, and then spread throughout the possible habitable regions in our neighborhood. It is also possible that life could have originated outside the solar system altogether, as suggested by Hoyle. In this scenario, known as *panspermia,* life is introduced to our world by a passing comet or asteroid.

Within our lifetimes, we stand a good chance of making progress on this vital issue. If future missions sent to Mars ultimately show that life once thrived on its now barren surface, we can determine whether Martian biology shares a common origin with life on Earth. It would also be interesting to dispatch a probe beneath the icy crust of Europa, a moon of Jupiter, and into the liquid ocean below. According to current scientific thought, the ocean water is warmed by tidal heating and can possibly be the home for some variety of life. The discovery of independently evolved life, on either Mars or Europa, would indicate that life is a likely outcome on any habitable planet. If these unearthly worlds turn up sterile, however, this issue is likely to remain unsettled for a long time to come.

We must distinguish between the Earthlike planets that develop primitive life only, and those that evolve intelligent life capable of interstellar communication. Only a fraction of life-bearing planets support an intelligent species, and only a fraction of these intelligent species develop interstellar communication. Because the advance from one-celled life forms to multicellular life required the longest time span on Earth, a bottleneck is likely to exist at this early stage. On the other hand, nothing indicates that a steady advance in the complexity of one-celled organisms does not lead inevitably to multicellular forms. For example, the Ediacara fauna may have been independent multicellular forerunners, whereas the plants, animals, and fungi of today emerged independently from single-celled ancestors. Taken together, these findings argue rather eloquently for the hypothesis of inevitability.

The final ingredient is the lifetime of an advanced alien civilization, compared to the lifetime of its parent star. For our planet Earth, this duty cycle is currently very small. Although the planet is 4.6 billion years old, our civilization has been leaking radio broadcasts into space for only about one century. In principle, however, our interstellar communication can continue into the far future of the planet. In the most optimistic scenario, Earth can broadcast radio signals until the Sun becomes a red giant and pasteurizes our world. In this case, Earth would provide a beacon of interstellar communication for about half of its total lifetime.

How can we make sense of the uncertainties in this discussion? The astronomical components are rapidly becoming much better known. With moderate confidence, we estimate that the galaxy contains roughly one billion habitable planets. The fraction of these hypothetical planets that contain intelligent life capable of interstellar communication, however, is far more the province of rampant speculation than of informed opinion. One of the largest uncertainties lies in the fraction of the habitable planets that actually develop life of any variety. The history of life on Earth suggests that once the evolution of life begins, the development of complexity, intelligence, and even technology is reasonably likely. At the risk of sounding hopelessly naive in hindsight, we suggest that roughly one in ten thousand of the habitable planets develop life. Folding together the remaining factors, and assuming (with little justification) that civilizations with technological capabilities do not readily destroy themselves, we estimate that the total number of civilizations in the galaxy is roughly one thousand. If these civilizations are randomly distributed throughout the galactic disk, the typical distance between neighboring civilizations is about three thousand light-years.

Although this optimistic appraisal suggests that the galaxy may contain a respectably large number of intelligent societies, the chances that we will soon make contact remain extremely remote. The isolation imposed by the enormous distances between stars is vast and profound. Suppose that the nearest intelligent civilization is three thousand light-years away. For comparison, our current census of nearby stars is complete out to a distance of only fifteen light-years; astronomers are continually finding new—previously undetected—red dwarfs at surprisingly close distances, from fifteen to thirty light-years. Within three thousand light-years, the typical distance to an alien civilization, millions of stars remain hidden, waiting to be discovered. A nearby alien civilization could be broadcasting as much energy as an entire star and still remain completely undetected. Of course, the energy requirements for interstellar communication decrease substantially if the civilization in question uses a directed signal, one that is beamed in a particular direction. That strategy has the shortcoming of only broadcasting to a small fraction of the sky. Thus, the bottom line is not very good. Although a thousand civilizations could conceivably live within our galaxy, a great deal of luck is necessary to establish contact.

COLONIZATION OF THE GALAXY

Given the likelihood of life in other solar systems, intelligent and otherwise, let's consider the possible colonization of the galaxy. We can make the distinction between natural processes and those directed by intelligence of some sort.

Even with no intelligence to direct it, life can spread throughout the galaxy through naturally occurring events. The mechanism to propagate life could be meteors or asteroids that collide with a life-bearing planet, fly off into space, and subsequently carry the seeds of life to a new planet in a new solar system. The directions of the meteors are determined purely by chance, and hence they fly off in random directions. Instead of traveling from point to point in a straight line, known to be the most direct route, the propagation of life takes random steps and eventually wanders away from the planet of origin. This process, called a *random walk*, is a rather inefficient mode of travel.

To estimate the time required to travel the length of the galaxy by a random walk, we can assume that the galaxy is thirty thousand light-years across, and the step length traveled by a life-carrying meteor is a few light-years, the typical distance between stars. The meteors or comets that carry life across the interstellar void travel at about 30 kilometers per second, the typical random speed of stars in the galactic disk (this random velocity is superimposed on the ordered flow of stars around the galactic center). In this scenario, the time required to randomly populate the galaxy with life is nearly three trillion years, about three hundred times longer than the current age of the universe. It is highly unlikely that life has spread throughout our galaxy through this mode of propagation. For comparison, the time required to evolve life spontaneously is much shorter, about four billion years here on Earth. Given the relative youth of the universe and the galaxy, spontaneous life appears to be much more likely than life propagated through random processes.

The propagation of life throughout the galaxy can also take place in a directed manner. Suppose that civilizations can produce transportation with speeds comparable to the meteors discussed above, about 30 kilometers per second. The travel time between stars is then about thirty thousand years. Since this time interval encompasses many generations, at least for humans, we expect that most of the time bottleneck occurs in the transportation time. As a result, an ambitious civilization can spread out through the galaxy in an advancing front with a steady speed. The time required to colonize the entire galaxy in this scenario is approximately the travel time necessary to cross the galaxy, about three hundred million years. Thus, the estimated time required to colonize the galaxy is somewhat shorter than the time to evolve intelligent life, four billion years, for the case of Earth.

Overlooking the inherent uncertainties in these estimates, we can summarize the situation regarding galactic colonization. A capable and purposeful civilization can in principle colonize the galaxy in about one billion years. With a random process of propagation, life is likely to require trillions of years to spread throughout the galaxy. For comparison, the evolution of intelligence takes about four billion years and the age of the galaxy is nearly ten billion years. Thus, the galaxy is already old enough that directed galactic colonization could have plausibly taken place. But it hasn't. No contacts of any kind have been made with extraterrestrial societies. This silence has some significance.

The most likely explanation is that no civilization has overcome the enormous barriers associated with interstellar travel. Actually traveling in spaceships is probably highly impractical, even for advanced societies. In order to propel a craft up to the requisite speeds, the energy requirements are extraordinary. Furthermore, it is not clear what the economic incentive of such travel would be. Because most stars are smaller than the Sun, and hence live longer than the current age of the universe, it is unlikely that an advanced civilization would be forced to leave its solar system by stellar evolution.

In the future, however, the Stelliferous Era unfolds in a manner more favorable to the propagation of life. Both the possibility and the motivation for interstellar travel increase dramatically. As relatively heavy stars like the Sun begin to burn out and turn into red giants and then white dwarfs, intelligent civilizations living around these dying stars might find it attractive to colonize the planetary systems of nearby stars which are still actively generating energy through nuclear fusion. For example, in several billion years, a civilization compelled to abandon a dying 1.05-solar-mass star might find an attractive, albeit temporary, home in our solar system. At this late date, Earth will be reduced to an unpleasant imitation of Venus, but Mars may well be beckoning with warm and wet conditions, as well as a freshly regenerated atmosphere and a bullish real estate market.

Over the longer term, conflicts of priority may well arise as advanced civilizations are routed from stars of progressively lower masses. For a long time to come, however, the number of viable planetary systems should steadily increase as the red dwarfs come on line and reheat their long-frigid planets. Only after the smallest red dwarfs, those with only 8 percent of a solar mass, have sparkled and faded, will planets with liquid water become largely a frozen memory. Although unavoidable, this energy crisis of the future will not take place for trillions of years.

The possibility of galactic conquest raises an interesting issue. If any one particular intelligent civilization turns out to be hostile and aggressive, then the galaxy is unstable to being plundered, pillaged, and conquered. The hostile civilization can simply take charge and take over. We have argued above that a sufficiently advanced and ambitious civilization could in principle colonize the entire galaxy within about one billion years. If an aggressive civilization does eventually evolve and solve the problems associated with space travel, our galaxy may well be conquered by it in the future.

An alternative possibility exists. A civilization sufficiently advanced to contemplate interstellar travel may realize the futility of the effort and might not bother to colonize other solar systems. Such a society would presumably also have a good understanding of the origin and evolution of life. If a civilization can figure out the fundamental basis of consciousness, it can transcend natural selection. For example, the ascent of life may be essentially automatic and life may arise naturally on any suitable planet. It may also turn out that evolution always produces intelligent species of roughly comparable quality. In this case, the life that arises naturally on any habitable planet would be "just as good" as that of any civilization that could migrate there. Once these concepts are understood, any thinking civilization would simply stay at home.

In making predictions regarding extraterrestrial intelligence and galactic colonization, we should keep in mind the precarious nature of this endeavor. Going back almost two centuries, in 1835, the well known French natural philosopher Auguste Comte wrote, "On the subject of stars, all investigations which are not ultimately reducible to simple visual observations are . . . necessarily denied to us. . . . We shall never be able by any means to study their chemical composition. . . . I regard any notion concerning the true mean temperature of the various stars as forever denied to us. . . ." In the glaring light of modern astronomy, this prediction seems hopelessly naive. Modern-day speculations concerning interstellar communication and space travel should be taken with an appropriate dose of caution. In contrast, predictions regarding the physical universe, especially the future evolution and death of the stars, rest on a much firmer foundation.

THE END OF STAR FORMATION IN THE GALAXY

Even the longest-lived stars die after trillions of years. Although this span of time seems long from a human perspective, and even compared to the current age of the universe, it is not very long in the broader context of the distant future. Because stars have finite lifetimes, galaxies can maintain their present status only as long as they continue to manufacture new stars. How long can a galaxy sustain normal star formation before it runs out of raw material? Which cosmological decade will preside over the last generations of stars?

The history of star formation in our galaxy, and in other spiral galaxies, is a complex subject. Although the future of star formation in the galaxy is more uncertain, we can provide a rough outline of the story. In order to create new stars, galaxies must convert gas into clouds, which provide the sites for star formation. Once enough gas has accumulated into the clouds, so that they are gravitationally bound together, the clouds naturally give birth to stars. Galaxies can continue to manufacture stars as long as they can make clouds, which requires the galaxy to have a reliable gas supply. Star formation ceases to function when a galaxy literally runs out of gas. We would thus like to know how long it takes for a galaxy to deplete its reservoirs of raw material.

To obtain a crude projection of the time left for star formation, we can divide the mass of the gas currently stored within the galaxy by the rate at which gas is converted into stars. This simple accounting procedure indicates that the galaxy will deplete its gas reserves in another ten billion years, roughly comparable to the current age of the universe. Clearly, galaxies cannot continue to produce stars indefinitely. However, the actual time scale for gas depletion is somewhat longer than this naive estimate.

When stars die, they return a portion of their mass back to the interstellar medium, the galactic storage facility for gas and dust. Stars like the Sun lose a substantial fraction of their mass through stellar winds and the ejection of planetary nebulae. Higher-mass stars return most of their mass to the interstellar medium when they explode in supernovae. The stars themselves thus provide a source for gas, which is then incorporated into future generations of stars. Additional gas is added to the galaxy as external material falls onto the galactic disk. Finally, and most importantly, the star formation rate grows smaller as the total gas supply decreases. These recycling effects and conservation procedures stave off the epoch when galaxies run out of gas and cease to produce new stars. Even with all of these complications, however, the galactic gas supply will be depleted within several trillion years. By the 14th cosmological decade, when the universe is one hundred trillion years old, essentially all of the conventional star formation in galaxies will grind to a halt.

Purely by coincidence, both star formation and stellar evolution draw to a close at approximately the same cosmological decade. Low-mass stars—red dwarfs—burn hydrogen and live productive stellar lives for trillions of years, roughly comparable to the time it takes galaxies to run out of gas and exhaust their star formation capabilities. The universe thus changes its character rather abruptly. During the 13th cosmological decade, stars are shining brightly and the universe is an energetic place. After the 14th cosmological decade, the stars have gone out and the universe will appear much darker to human eyes.

3

THE DEGENERATE ERA

$15 < \eta < 39$

DEAD STELLAR REMNANTS CAPTURE DARK MATTER, COLLIDE WITH EACH OTHER, SCATTER INTO SPACE, AND FINALLY DECAY INTO NOTHINGNESS.

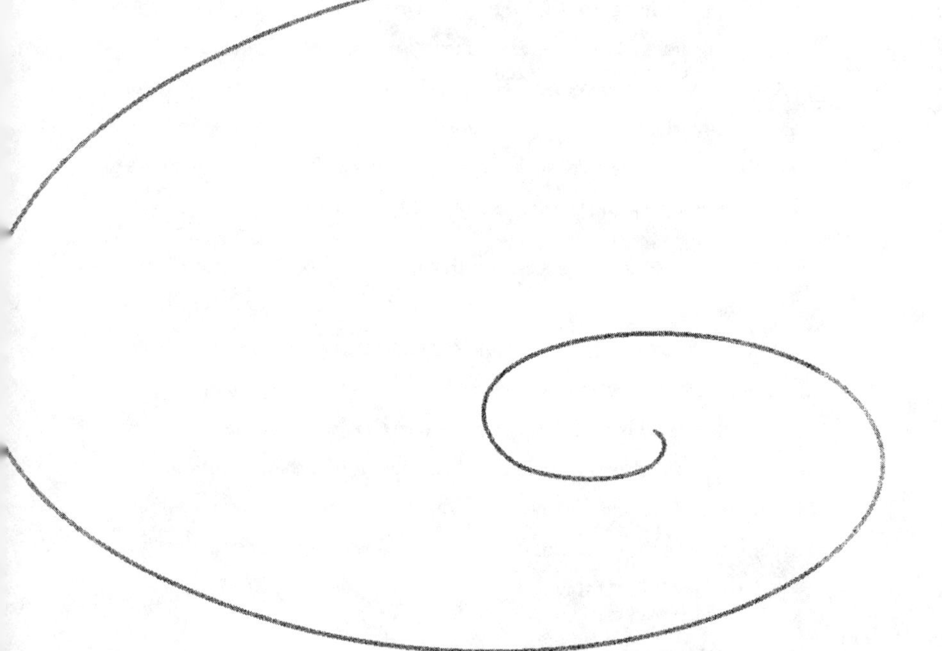

Cosmological decade 15, near the surface of a white dwarf:

Miranda gazed out the porthole of the spacecraft to view her world for one final time. As the launch sequence began, she was both sad and excited about the prospects of leaving this established civilization, and setting out to find a new location for a colony. The spherical metallic platform, which spread out beneath her, was so flat that its surface curvature could barely be discerned. This immense structure, with faintly glowing cities and artificial landscapes, had sustained her ancestors for countless generations.

The metallic surface supporting the colony almost completely enclosed the crystallized white dwarf. The structure had been designed with exquisite precision to capture the sparse radiative energy generated by this remnant of a long dead star. Through the capture and annihilation of dark matter, a naturally occurring process, the white dwarf produced enough energy to support one billion citizens. Now, however, the population had grown and more resources were needed. It was time to find a new place to live.

In a reflective mood, Miranda imagined what it would have been like in the distant past, when rich clouds of hydrogen gave birth to bright young stars. How different it must have looked, with billions of stars lighting up the sky in every galaxy. But that profligate universe of the past was now long dead. How can someone who lives only a few hundred years ever fully comprehend time scales of trillions of years? She closed her eyes to ponder this mystery, and the spacecraft gently lifted off.

Meanwhile, just below the surface of the white dwarf, seemingly innocuous events of great import were taking place. With a tortuously slow rate of progress, imperceptible to the high-temperature beings living outside, chemical reactions were gradually assembling large molecules into ever longer structures. This increase in complexity was powered by occasional bursts of high-energy radiation leaking out of the stellar interior. Even as Miranda and her kind tenaciously clung to existence in an increasingly inhospitable universe, the building blocks for a new type of biology were being synthesized for the first time.

What happens when the stars have stopped shining? In a hundred trillion years, the last generations of stars will have been wrung from depleted interstellar clouds, and the evolution of the few remaining red dwarfs will be slowly drifting to a finish. As the dynamic cycle of stellar births and deaths fades into a memory, the universe shifts its temperament, restocks its inventory, and continues its evolution.

As the universe enters the *Degenerate Era,* the effects of change are quite evident. Conventional hydrogen-burning stars have given way to stellar remnants: brown dwarfs, white dwarfs, neutron stars, and black holes. Although these objects may seem cold and desolate, they will be the source of action and excitement in the universe. The clock pacing the rate of unfolding events runs much more slowly. Astrophysical events that could never happen in the universe of today, because of severe time limitations, begin to occur.

INTRODUCING THE DEGENERATE STELLAR REMNANTS

A stockpile of stellar remnants provides the astronomical grist of the Degenerate Era. We have already met this cast of degenerates in the previous chapter. In this collection of stellar remnants, the four usual classes—brown dwarfs, white dwarfs, neutron stars, and black holes—are the net result of trillions of years of stellar evolution (see the following figure). For the sake of completeness, however, we should remember that a fifth possible fate exists. When instabilities ensue within a sufficiently massive conventional star, the ensuing supernova explosion can sometimes be so violent that every shred of stellar matter is dispersed into space. In other words, *nothing is left.* This outcome represents an early and decisive victory for thermodynamics in its battle with the force of gravity. In the other four cases, gravity puts up a far more tenacious effort.

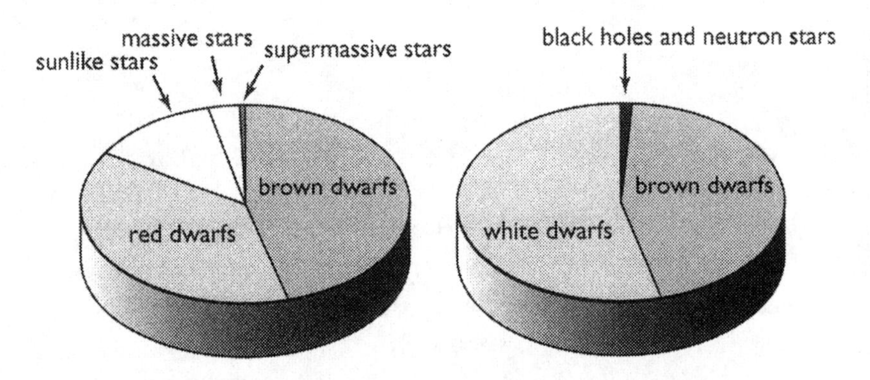

The pie chart on the left shows the relative number of stars born in different ranges of mass. The largest sector is for brown dwarfs, which have masses ranging from 0.01 to 0.08 solar masses. The other large sector shows the red dwarfs, which have masses lying between 0.08 and 0.43 solar masses. The next largest sector contains the middleweight stars, which range between 0.43 and 1.2 solar masses. Massive stars fall between 1.2 and 8 solar masses, but the smallest slice is reserved for the heavyweight stars, which are heavier than 8 solar masses. The pie chart on the right shows the distribution of stellar remnants, the objects left over after stellar evolution has run its course. The brown dwarfs remain as brown dwarfs, but most stars (those with less than 8 solar masses) end their lives as white dwarfs. Only the tiny fraction of stars heavier than 8 solar masses can become black holes and neutron stars. The size of the black hole and neutron star slice has been exaggerated for clarity.

Brown Dwarfs

Brown dwarfs are larger than planets, smaller than regular stars, and represent the lightest type of degenerate remnant. They are stellar failures in the sense that they are unable to achieve nuclear ignition of hydrogen in their interiors. They have no access to the usual source of stellar energy, and so from their moment of birth forward, they are consigned to a low-profile life of cooling and contraction.

Several physical reasons underlie the failure of brown dwarfs to become stars. An important part of the explanation is that nuclear reaction rates are extremely sensitive to changes in temperature. Inside a star, a slight increase in temperature produces a gigantic surge in the power generated by hydrogen fusion. As a result, the temperature at which hydrogen fusion occurs in stars is always close to ten million degrees kelvin. (Should the core of a star grow hotter, the surge of excess energy causes it to expand and cool.) Next, because the temperature is fixed at ten million degrees, the density of a stellar interior goes up as the mass of the star goes down. Small stars must contract more drastically in order to achieve a central temperature of ten million degrees, and so smaller stars are considerably denser than more massive stars. The final ingredient is that the pressure produced by degeneracy increases rapidly with increasing density. That is, if you try to squeeze a chunk of degenerate material, it is very stiff and resistant to compression.

By linking these phenomena together, we can see why stars must contain more than a certain minimum mass in order to burn hydrogen. As the mass of the star gets smaller, the central density gets larger. But, if the central density becomes too large, degeneracy pressure dominates over ordinary thermal pressure and supports the star before the temperature reaches the requisite ten million degrees. The onset of degeneracy pressure thus enforces a maximum central temperature that is attainable by a star with a given mass. Sufficiently small stars have a maximum temperature lower than ten million degrees, the hydrogen burning temperature. If an aspiring stellar object contains too little mass, it cannot burn hydrogen, and will never become a full-fledged star.

The smallest stars capable of sustained nuclear fusion have about 8 percent of the mass of the Sun. Stellar objects falling below this minimum mass are brown dwarfs. The radial size of a brown dwarf is roughly comparable to an ordinary small star, about one tenth the size of the Sun, or about ten times the size of Earth. A final relevant characteristic of brown dwarfs is their chemical composition. Because they don't actually *do* anything, these substellar slackers retain precisely the elemental abundance they are born with. Consequently, they are primarily composed of hydrogen.

In the past few years, astronomers have detected a growing number of brown dwarfs, and indeed, brown dwarfs are expected to be very abundant. A galaxy the size of the Milky Way probably contains billions of brown dwarfs. And although brown dwarfs have not had much influence on the cosmos thus far, these failed stars will have their day as the universe grows older. During the Degenerate Era, brown dwarfs will contain most of the unburned hydrogen left in the universe.

White Dwarfs

The vast majority of stars, including our own Sun, end their lives as white dwarfs. Although a dim 0.08-solar-mass star is one hundred times lighter than a tempestuous 8-solar-mass star that shines with the light of three thousand Suns, both are destined to finish their evolution by becoming white dwarfs. At the close of the Stelliferous Era, our galaxy will contain nearly one trillion white dwarfs, with a comparable number of brown dwarfs. The white dwarfs are individually much more massive and will contain most of the ordinary baryonic mass in the universe.

White dwarfs have a range of masses, with the average somewhat less than a solar mass. The smallest progenitor stars lose very little of their mass as they evolve to become white dwarfs. A small red dwarf star ends its life as a white dwarf with nearly its original mass. Stars like the Sun, which are destined to swell into red giants, lose a considerably larger fraction of their initial mass. The Sun will give rise to a white dwarf with about 0.6 solar masses. The larger stars, on the other hand, shed most of their mass in becoming white dwarfs. An 8-solar-mass star, for example, ends its life as a 1.4-solar-mass white dwarf. The remaining mass is blown off in a wind during the red giant phase and is returned to the interstellar medium for recycling.

The white dwarfs that we see in the skies today are those in the upper half of the possible mass range. Because of the relative youth of the universe and its stellar content, only stars larger than about 0.8 solar masses have had time enough to die. Smaller stars are more numerous and live much longer. The smallest stars, near the minimum of 0.08 solar masses, have barely begun their evolution. In the future, however, even the smallest stars will burn out and become white dwarfs. By the beginning of the Degenerate Era, the most common white dwarfs will have relatively low masses.

A white dwarf with a typical mass of 0.25 solar masses has a radius of 14,000 kilometers, about twice the size of Earth. Curiously, heavier white dwarfs are actually smaller in size. A one-solar-mass white dwarf has a radius of only 8700 kilometers. White dwarfs have this strange property—more massive objects are smaller—because their matter is degenerate. This odd behavior is in direct contrast to ordinary matter. If you add mass to a rock, it gets bigger. If you add mass to a white dwarf, it will shrink!

Why can white dwarfs be seen at all? If white dwarfs are the end result of stellar evolution, after nuclear fusion processes have shut down, why do these stars shine? These stellar remnants contain a huge supply of heat energy left over from their fiery past. This enormous reservoir of heat radiates away into space at a painstakingly slow pace. As a result, we can see white dwarfs in the sky. As the stars get older, they grow cooler and dimmer, much like the dying embers from a fire. The time required for a white dwarf to cool down is billions of years, comparable to the age of the universe today. When the universe enters the Degenerate Era trillions of years from now, white dwarfs will attain the frigid temperature of liquid nitrogen. Further cooling is arrested by an unusual internal energy source that we will meet later in this chapter.

The curious fact that white dwarfs of smaller mass are larger in size poses yet another question. What happens as we make the mass of a degenerate remnant smaller and smaller? Does the object just keep getting bigger? There is a limit. As the stellar mass gets smaller and the size grows larger, the density of the material decreases. As the density decreases below a critical level, the matter is no longer degenerate and ceases to behave in this counterintuitive manner. When a star has too little mass to be degenerate, it acts like ordinary matter. A stellarlike object thus has a minimum mass required in order to be degenerate. This mass is approximately one-thousandth of the mass of the Sun, about the same mass as Jupiter. Lightweight objects with less than one-thousandth of a solar mass do not behave in a degenerate fashion. They act like ordinary matter and are known as planets.

On the other hand, white dwarfs cannot be made too massive. An overweight white dwarf will explode in dramatic fashion. As a white dwarf becomes heavier, it becomes smaller and denser, and a larger pressure is required to hold up the star against the opposing force of gravity. To maintain a larger pressure, in this case degeneracy pressure of the electrons, the particles must move faster. When the density is so great that the requisite particle speed approaches the speed of light, the star runs into deep trouble. Einstein's theory of relativity places a strictly enforced speed limit—no particles are allowed to go faster than light. When a star reaches this state where particle speeds faster than light are necessary, the star is doomed. Gravity overwhelms the degeneracy pressure, instigates a catastrophic collapse, and thereby initiates a stellar explosion—a supernova. These spectacular explosions are comparable in magnitude to those that mark the passing of massive stars (as discussed in the previous chapter).

To avoid the fiery death of a supernova, a white dwarf must have less than 1.4 solar masses of material. This vitally important mass scale is called the *Chandrasekhar mass,* appropriately named after the eminent astrophysicist S. Chandrasekhar. At age 19, he worked out this mass limit while on an ocean passage from India to Britain, prior to starting his graduate studies at Cambridge University in the 1930s. He subsequently won the Nobel Prize in physics for his contributions to astrophysics.

Neutron Stars

Although white dwarfs are incredibly dense, a *neutron star* embodies an even more compact form of stellar matter. The typical density of a white dwarf is "only" about one million times the density of water. The nuclei of atoms are much denser, however, about a quadrillion (10^{15}) times denser than water, or a billion times denser than a white dwarf. If a star can be squeezed to the enormously high density of an atomic nucleus, the stellar material can attain an exotic but stable configuration. At these high densities, electrons and protons prefer to live together in the form of neutrons so that essentially all of the matter resides in neutrons. The neutrons become degenerate and the pressure they produce, again due to the uncertainty principle, supports the star against gravity. The resulting neutron star is very much like a single gigantic atomic nucleus.

The unfathomably high densities required to produce a neutron star are naturally achieved during the collapse of a massive star at the end of its life. Within a highly evolved star, the central region becomes a degenerate iron core that collapses to initiate a supernova, often leaving behind a neutron star as a remnant. Additional neutron stars can be produced by the collapse of white dwarfs. If a white dwarf gains mass slowly, from a companion star for example, it can sometimes avoid a supernova explosion and contract to become a neutron star.

Compared to white dwarfs and brown dwarfs, neutron stars are relatively rare. Only those stars born with masses greater than eight times that of the Sun can explode to create neutron stars when they die. These massive stars represent only the high-mass tail of the distribution of stellar masses. The vast majority of stars are too small. Only about one out of four hundred stars are born large enough to achieve detonation and leave behind a neutron star. Despite these poor odds, a respectably large galaxy will contain millions of neutron stars.

The typical mass of a neutron star is about 1.5 times that of the Sun. Just as for white dwarfs supported by electron degeneracy pressure, neutron degeneracy pressure cannot support a stellar remnant of arbitrarily high mass. If the mass becomes too great, gravity wins its battle against degeneracy pressure and the star must collapse. The largest possible mass of a neutron star is somewhere between two and three solar masses, but is not precisely known. The properties of matter become highly exotic and somewhat uncertain at the mind-boggling densities achieved in the center of a neutron star. In spite of being heavier than the Sun, neutron stars are rather small, only about 10 kilometers in radius. A small size combined with a large mass implies an enormous density. A cubic centimeter of neutron star material, the size of a sugar cube, has the mass of nearly a billion elephants!

Black Holes

The fourth possible fate of a dying star is to end up as a black hole. After they explode and expire, the most massive stars may leave behind a remnant larger than the maximum mass for a neutron star, between two and three solar masses. A sufficiently massive stellar remnant cannot be supported by degeneracy pressure and must collapse to become a black hole. In a similar vein, fully formed white dwarfs and neutron stars can gain additional mass, usually from binary companions, and grow too large to be supported by degeneracy pressure. The resulting overweight remnants also must collapse and sometimes can produce black holes.

Black holes are strange beasts, with gravitational fields so strong that light itself cannot escape. This property is in fact the defining characteristic of black holes. For these objects, the escape velocity, that required to leave the surface, is larger than the speed of light. Because of Einstein's relativistic speed limit—nothing travels faster than the speed of light—no particles or radiation are allowed to leave a black hole. This apparently firm statement is not completely true, however, because of Heisenberg's uncertainty principle. In the very long term, black holes must eventually surrender their tightly held masses, but not until long after the Degenerate Era has ended.

Black holes are incredibly compact. A black hole with the mass of the Sun is only a couple of kilometers, about one mile, in radius. As another benchmark, a black hole the size of a baseball has about five times the mass of Earth. These singular stellar sensations exhibit a rich variety of other exotic properties, which we explore in the following chapter.

Massive stars are relatively rare, and the black holes they produce are even rarer. Less than one star in three thousand has a chance to become a black hole after its nuclear burning life is over. Because of their scarcity, these stellar understudies don't play a leading role until after the Degenerate Era has drawn to a close.

In addition to black holes resulting from stellar death, another variety of black holes inhabits our universe. Black holes in this second class are found at the centers of galaxies. Compared to their stellar counterparts, these *supermassive black holes* are truly enormous. They weigh in anywhere from one million to a few billion solar masses. As a point of reference, the effective radius of a black hole with a million solar masses is roughly four times the radius of the Sun.

GALAXIES IN COLLISION

Our galaxy, the Milky Way, now contains a hundred billion shining stars, which are collectively visible as a faintly glowing band that spans the night sky. In the Degenerate Era, the sky will appear pitch black. But the largest galaxies, held together by the gravity of cold dead stars and dark matter, will remain intact.

The most imminent threat facing ordinary galaxies like the Milky Way is not the death of their constituent stars, but rather a disruptive collision with another galaxy. Galaxies usually live in clusters or groups. Clusters are held together by gravity, and each galaxy traces its own orbit through the cluster. When large, loosely knit objects like galaxies pass by one another, they experience a kind of friction which causes them to sink toward the center of the cluster. Near the cluster center, galaxies are relatively closely packed, and they tend to collide with one another.

Galactic collisions will affect the universe of the relatively near future. Some galaxies are even colliding during the current era, the Stelliferous. As the universe fades into the Degenerate Era, the implications of these galactic interactions will become increasingly important.

When galaxies collide, the stars belonging to the two original galaxies intermingle to form a larger, but more disorganized, composite galaxy. Instead of displaying the elegant spiral patterns of solitary disk galaxies, a merged composite galaxy is jumbled and amorphous. During a collision, long streamers of stars, also called tidal tails, are ripped out of the galaxy. The orbits of the stars become complicated and irregular. A merged galaxy is a mess.

Galactic collisions are frequently associated with powerful bursts of star formation. Large clouds of gas within the galaxies merge during such collisions and produce new stars at prodigious rates. The multiple supernovae resulting from the deaths of the more massive stars can have quite dramatic consequences.

Although the structure of a galaxy as a whole appears completely different after a collision, the individual stars and their solar systems experience almost no effects. A galaxy like the Milky Way is mostly empty space—the stars in a galaxy are like individual grains of sand separated by several miles in every direction. Even within slightly denser merged galaxies, the star-to-star separation is more than a light-year, a thousand times larger than the solar system, and ten million times larger than a star. The planetary systems within a colliding galaxy are completely unperturbed by the slow-motion catastrophe taking place in the background and unfolding over millions of years. The most noticeable consequence for an Earthlike planet would be the gradual doubling of the number of stars visible in the night sky.

Indeed, the Milky Way is scheduled to experience a galactic collision—and lose its separate identity—in the relatively near future. The neighboring Andromeda galaxy, also known as M31, is presently directed on a collision course with the Milky Way. Because of the difficulty in making precise astronomical measurements of galactic velocities, we are not certain of the exact direction Andromeda is headed. It is clear, however, that this large galaxy will pass close by our own galaxy, and perhaps even collide with it, in about six billion years, just as the Sun is starting to swell into a red giant. Even if Andromeda and the Milky Way escape a collision in this particular future encounter, they cannot avoid each other in the long run. The Milky Way is clearly gravitationally bound to Andromeda. As the two galaxies orbit around each other, and energy is lost to dynamical friction, a future merger becomes almost inevitable.

The long-term fate of galaxy clusters is thus sealed: The member galaxies will eventually interact and merge. Their separate galactic identities will be subsumed as the entire cluster coalesces into one large and disorganized collection of stars. As the universe crosses over from the Stelliferous to the Degenerate Era, the clusters of today will become the large galaxies of the future. Indeed, our entire local group of galaxies, including the Milky Way and Andromeda, will be forged into a single metagalaxy.

GALAXIES IN RELAXATION

The gulfs between the stars in a galaxy like the Milky Way are so large that very few, if any, direct collisions between stars have ever taken place. So far. Continuing a familiar theme, rare events can occur if given enough time. As the Degenerate Era gets underway, collisions and near collisions between stars become increasingly important. These encounters will completely rearrange the structure of the galaxy and ultimately lead to its demise. Since this epoch of destruction will not arrive until well into the Degenerate Era, however, the stars will be stellar remnants and the galaxy will have long since been folded into the sprawling product of successive galactic mergers.

Even in the Degenerate Era, direct head-on collisions between stars are relatively rare. Close encounters and near misses are much more frequent than true collisions. As the Degenerate Era unfolds, stars pass near each other on a regular basis, and interact through their joint gravitational attraction. The close passage of two stars leads to a small adjustment in the speed and direction of each star. The stars tend to scatter off of each other whenever they draw near, as illustrated in the figure on the next page.

Over the course of time, many scatterings take place, and their effects slowly accumulate. The net result of the long sequence of scattering events is to redistribute the individual speeds of the stars orbiting within the galaxy. The smaller and lighter stars tend to pick up speed and gain orbital energy, whereas the heavier stars tend to lose their orbital energy. When a large number of stars participate in this redistribution of wealth, the galactic structure slowly changes in a process of *dynamical relaxation.* As this relaxation proceeds, some stellar remnants are given so much energy that they must leave the galaxy. As time goes on, more and more stars evaporate from the dying galaxy, and are sent hurtling into intergalactic space at speeds of 300 kilometers per second (675, 000 miles per hour).

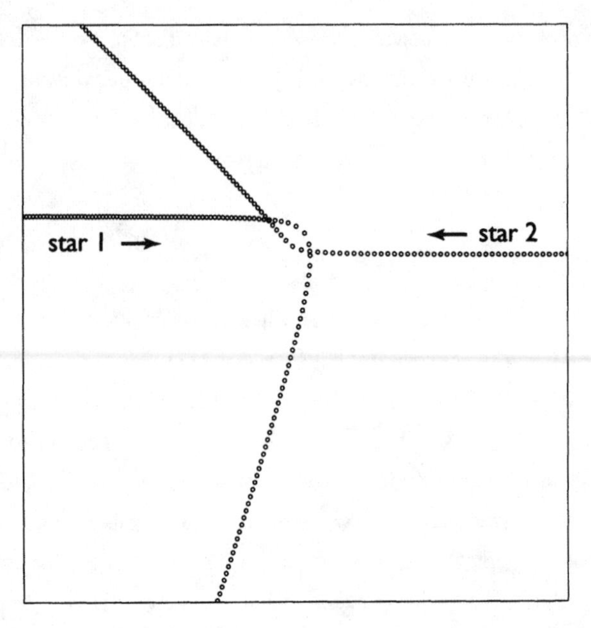

This diagram shows the response of two stars to a close encounter. After the interaction, each star has a new direction, energy, and hence velocity. A very large number of these encounters will dynamically relax the galaxy and thereby change its structure over vast expanses of time.

As dynamical relaxation runs its course, the number of exiled stars increases, and major structural changes are forced upon the galaxy. Since the stars that leave have the highest energies, the stars left behind have less energy on average. Energy is thus drained away. In response to the mounting energy crisis, the galaxy is compelled to grow smaller and denser. This galactic downsizing instigates even more stellar encounters and the banishment of more stars. As this process accelerates, a runaway situation can develop as the galaxy ejects the majority of its stars, and leaves behind a small minority in a tightly bound clump.

An unpleasant prospect awaits the low-energy stars that sink to the galactic center, where nearly every galaxy is thought to harbor a supermassive black hole. These enormous black holes contain millions to perhaps billions of solar masses. As a galaxy relaxes, the central black hole absorbs wandering stars that stray too close, within the event horizon. Throughout the Degenerate Era, these supermassive black holes will slowly gain weight on their steady diet of windfall stars.

Galaxies will persist for billions of times longer than the present age of the universe. Their remarkable durability is made possible by the huge separations between individual stars, and the slow rate at which stars cross those distances. But given enough time, galaxies must face their demise. During the next 19 to 20 cosmological decades, 10^{19} to 10^{20} years, most of the dead stars in a galaxy will escape through the process of stellar evaporation. A small but unfortunate fraction of the stars, perhaps 1 percent, will be consumed by the central black hole. After this process of dynamical relaxation has run its course, the life of a galaxy effectively comes to an end.

As the galaxy relaxes and disperses, the close encounters of passing stars have a devastating impact on any planets still circling the stars. These stellar scattering events tend to dislodge planets from their orbits and send them careening into the void. The fate of such a homeless planet was outlined in the previous chapter. Planets with orbital radii comparable to that of our Earth are scattered out of their solar systems during the 15th cosmological decade. Outer planets with larger orbits are more susceptible, and hence are long gone by this time. A planet like Neptune, with an orbital radius of 30 astronomical units, will be exiled in only 12 cosmological decades, a trillion years. Even the innermost planets can be removed from their orbits during the Degenerate Era. A planet with an orbit ten times smaller than that of Earth, somewhat smaller than the orbit of Mercury, will be displaced in about 17 cosmological decades. Thus, by the time they leave the galaxy during the 19th and 20th cosmological decades, stars will have long since been stripped of their solar systems.

The long-term future of planets in general, and our Earth in particular, is thus rather bleak. During the near term, planets will be pummeled by comets and asteroids, which cause global climatic changes and general cataclysmic destruction. Afterwards, the inner planets will be scorched and sterilized when their parent stars swell to red giant proportions. Any surviving planets will then be forcibly removed from their solar systems and sent off alone into the darkness of interstellar space.

DEGENERATE STELLAR COLLISIONS

The rare direct collisions between dead stellar remnants provide moments of extraordinary excitement, like exclamation points punctuating the almost endlessly desolate stretches of the Degenerate Era. These collisions can produce ordinary new stars, strange new types of stars, and spectacular explosions.

During this future epoch, most of the ordinary baryonic mass in the galaxy resides within white dwarfs. Although less mass is stored in brown dwarfs, which have smaller masses, roughly equal numbers are present. Within a large galaxy like the Milky Way, the population of both white dwarfs and brown dwarfs should number in the billions. As the dead stars trace through their orbits, direct collisions sporadically take place, with about one such smashing event every few hundred billion years. Given the current age of the galaxy, about ten billion years, the chances are good (roughly nine out of ten) that no stellar collisions have occurred so far. As the universe grows older than several hundred billion years, collisions will begin to add up. During the 15th cosmological decade, hundreds to thousands of stellar collisions will rock the galaxy.

The collisions between two brown dwarfs are of immediate astronomical, geological, and perhaps even biological interest. A large portion of the remaining hydrogen in the universe is locked up in brown dwarfs, which do not burn their hydrogen into heavier elements. When two brown dwarfs collide at a sufficiently direct angle, they can form a composite stellar object containing most of the original mass of the two stars (see the figure opposite). If its combined mass is above the threshold for stardom, the interaction product can contract and heat up until sustained hydrogen fusion lights up the newly formed stellar core. The star will turn on. The small red stars that emerge from these freak collisions will subsequently live for trillions of years.

Through these astronomical crashes, new stars can be made long after the interstellar medium has run out of gas. In a galaxy the size of the Milky Way, about a hundred of these stars will be shining at any given time. The combined starlight of these dim red stragglers endows the galaxy with a total power output comparable to that of the present-day Sun.

Brown dwarf collisions can also create new planets. Unless the collision

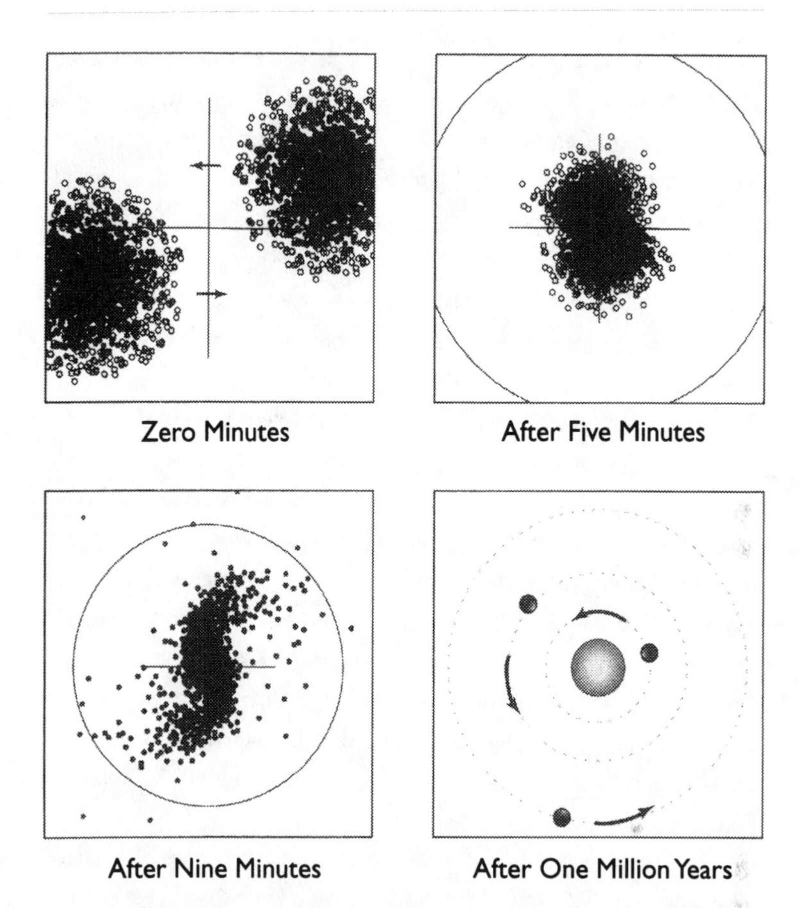

This computer simulation shows the collision between two brown dwarfs. The first three panels illustrate the first few minutes of the event. The final result of the collision, shown schematically in the fourth panel, is a true star with enough mass to initiate the fusion of hydrogen. The collision naturally produces a disk of gas and dust surrounding the newborn star; the disk is an environment in which planets are likely to form.

is precisely head-on, some of the gas from the brown dwarfs will be swirling too rapidly to be incorporated into the newly forged star. This spinning material readily forms a circumstellar disk of gas and dust around the young stellar object. Since the formation of planets is a likely outcome of an evolving disk, these new stars are thus inclined to give birth to new solar systems.

The planets that spring from the aftermath of a brown dwarf collision should contain all of the ingredients necessary for the evolution of life. A planet in the care of a red dwarf can stay warm for trillions of years, far longer than the current age of Earth. These systems contain an ample supply of heavy elements, including the carbon and oxygen that provide the backbone of terrestrial life. Liquid water may be available on planets with favorable orbits. In principle, familiar types of life can arise and evolve on such new planets as long as the galaxy remains together. Only after the 20th cosmological decade, when the galaxy evaporates and the brown dwarf collision rate vanishes, will the last Earthlike worlds fall victim to the eternal night.

Collisions between white dwarfs can lead to even brighter, albeit shorter-lived, fireworks. If two white dwarfs collide and merge, and their combined mass is over the Chandrasekhar limit, degeneracy pressure cannot support the merger product against gravitational collapse. The newly formed but overweight star must then explode as a supernova. About one out of ten collisions between white dwarfs will detonate a supernova. As a result, the galaxy is slated to experience about one explosion every trillion years as long as it remains intact, for about twenty cosmological decades. Supernovae are impressive today, but they will be truly spectacular in the impoverished environment of a dying galaxy during the Degenerate Era.

The most likely outcome of a rare collision between two white dwarfs is not a supernova, but rather the formation of an odd new type of star. Most white dwarfs descend from low-mass stars, and are composed of almost pure helium. A collision between two of these typical white dwarfs produces a somewhat larger stellar object made of helium. If the final collision product has a mass greater than 0.3 solar masses, the object can in principle ignite helium in its interior. These stars can fuse helium into heavier elements in an analogous fashion to evolved (old) stars of higher mass (as described in the previous chapter). In order for the star to burn helium, however, the collision must impart enough heat energy into the star, in much the same way that heat from a match is required to ignite a sheet of paper. If the star is too cool to burn its helium, it will contract to become just another white dwarf, wandering about the galaxy and waiting to either collide again or scatter into intergalactic space.

Compared to their conventional hydrogen-burning counterparts, these helium-burning stars are hotter, brighter, denser, and shorter-lived. A typical star, with one half of a solar mass, has a radius ten times smaller than that of the Sun and a luminosity ten times larger. The stellar surface is scorching hot, with a temperature of 35,000 degrees kelvin, about six times hotter than the Sun. The conditions in the stellar core are even more extreme, with temperatures of 100 million (10^8) degrees and a central density of nearly 10,000 grams per cubic centimeter. These stars live for only a few hundred million years, a long time by human standards, but a mere blink of an eye compared to their long formation time. Even if these stars produce planetary systems, they probably will not live long enough to see complex life evolve. If we extrapolate from the time line experienced by Earth, life in these systems is unlikely to evolve beyond the most primitive stages, as represented by viruses and unicellular biota.

When slightly heavier white dwarfs collide, another strange new type of star can be created. If the collision product is heavier than 0.9 solar masses, but smaller than the Chandrasekhar mass (so it won't explode), the new object can, in principle, sustain carbon fusion in its central core. The properties of a carbon-burning star are even more exotic than a helium star. A solar-mass carbon star is about one thousand times brighter than the Sun and its surface seethes at 140,000 degrees kelvin. By stellar standards, the radius of the star is tiny, only a bit larger than that of Earth. Inside the stellar core, the temperature approaches one billion degrees and the density is one hundred thousand times that of solid rock. These brightly burning candles only last for a million years. Any accompanying planets will still be in their early formative phases when the star depletes its nuclear fuel and turns off. Even the simplest biosphere is unlikely to find time to gain a foothold.

ANNIHILATION OF DARK MATTER

Galactic halos are composed largely of dark matter, much of which may reside in the form of nonbaryonic particles. Recall that baryonic material is primarily composed of protons and neutrons, and hence comprises most of what we consider to be ordinary matter. As we discussed in Chapter 1, current astronomical wisdom suggests that a large fraction of the mass in the universe must assume a nonbaryonic identity. And a great deal of this nonconventional matter is thought to reside in the halos of galaxies.

One of the leading candidates for the dark matter has been named *weakly interacting massive particles.* These rather strange particles, weighing in at ten to one hundred times the mass of the proton, interact only through the weak nuclear force and gravity. They have no electric charge, and thus are impervious to the electromagnetic force. They also have no interactions via the strong force, and thus cannot be bound into nuclei. Because these particles interact very weakly, they can live for a long time in diffuse regions such as the galactic halo. In particular, they can live much longer than the current age of the universe. Over sufficiently long time scales, however, these particles interact with ordinary matter and annihilate with each other.

The annihilation of dark matter takes place in two different settings. In the first case, when two particles meet in the galactic halo, they can interact and directly annihilate each other. In the second case, the particles are captured by stellar remnants, such as white dwarfs, and then subsequently annihilate with each other inside the stellar cores. Both of these mechanisms for decimation play an important role in the future of the galaxy and the universe.

In the galactic halo, dark matter particles have a low density, about one particle per cubic centimeter, and have rather large velocities, about 200 kilometers per second. Since the particles interact only through the weak force, the chances of annihilation are exceedingly small. After about 23 cosmological decades (10^{23} years), however, the halo population of dark matter will change drastically due to these interactions. When dark matter particles annihilate, they generally leave behind smaller particles with highly relativistic speeds, large enough to escape the gravitational pull of the galaxy. The end result of the annihilation process is thus to radiate away the mass energy of the galactic halo into intergalactic space.

Since dark matter accounts for a large fraction of the mass budget of the universe, the annihilation products from dark matter interactions provide an important part of the inventory of the universe at late times, especially between the 20th and 40th cosmological decades. The residual products from direct annihilation events in galactic halos provide a rich variety of particles, including photons, neutrinos, electrons, positrons, protons, and antiprotons.

Dark matter is captured by stellar remnants such as white dwarfs. The dark matter of the galactic halo provides a background sea of particles that are continually streaming through space. These particles also stream through all of the objects in the galaxy, including stars, planets, and, at the present cosmological time, people. Approximately one hundred billion (10^{11}) such particles are passing through you every second. Because the particles interact only through the weak force, and because the weak force is *very* weak, these particles pass right through all types of matter with essentially no effect. Every once in a while, however, a dark matter particle does interact with the nucleus of an atom and thereby relinquishes some of its energy.

If an interaction occurs within a white dwarf, the dark matter particle can remain gravitationally bound to the star. Over vast expanses of time, the population of these dark matter particles gradually builds up inside the object. The time required for dark matter to be captured through this process is much longer than the nuclear burning lifetimes of the stars, which live as stellar remnants for nearly all of this time. As the concentration of dark matter particles increases in the stellar core, the opportunities for particles to annihilate increase. In the long run, the star reaches a steady state in which annihilation occurs within the stellar remnant at the same rate that particles are captured from the galactic halo.

The process of dark matter capture and annihilation provides a vitally important energy source for the white dwarfs of the future. These stellar remnants are left over from stellar death, after nuclear reactions have shut down. In the absence of an additional energy source, white dwarfs would become colder and dimmer until they reach the background temperature of the universe. Because of the energy they usurp from dark matter annihilation, however, white dwarfs can radiate energy for a very long time. The total power produced by a single white dwarf through this annihilation process is about one quadrillion (10^{15}) watts. Although this feeble power output is about one hundred billion (10^{11}) times smaller than that of the Sun, this energy production mechanism will run the universe in the future. Such energy generation can continue as long as the galactic halo remains intact, roughly twenty cosmological decades (10^{20} years), or ten billion times longer than the hydrogen burning lifetime of the Sun.

The dark matter particles captured by white dwarfs ultimately annihilate into radiation, which eventually dominates the background radiation field of the universe. Before leaving the star, however, this radiation is degraded to longer wavelengths and hence lower average energies. Photons leave the stellar surface with a characteristic wavelength of about 50 microns (one-twentieth of a millimeter), one hundred times longer in wavelength than the light emitted by the Sun. This radiation is invisible to human eyes, but these infrared photons are easily detected using present-day technology. The surface temperature of the star is a cool 63 degrees kelvin, just below the temperature of liquid nitrogen.

At this epoch in the future history of the universe, galaxies will appear quite different than they do today. A typical galaxy contains billions of stellar remnants, all radiating energy through the process of dark matter capture and annihilation. An entire galaxy of these stellar remnants has a total power output comparable to that of our Sun. Scattered among these faintly glowing remnants are about one hundred more traditional stars manufactured through brown dwarf collisions. Although rather dim by the standards of today, these small stars will be veritable beacons of light against the darkness of the future. The total power generated by these few bona fide stars will outshine the billions of white dwarfs.

LIFE IN WHITE DWARF ATMOSPHERES

Although familiar types of life may well find their survival in jeopardy, an intriguing possibility for future life exists in the atmospheres of old white dwarfs. Any discussion of future life forms necessarily brings us into the realm of speculation. Besides being intrinsically interesting, however, the following chain of propositions cleanly illustrates the physical conditions inside white dwarfs at times in the far future.

After the death of its progenitor star, a white dwarf rapidly cools until its principal energy source is the capture and subsequent annihilation of dark matter particles. The white dwarf then remains in a more or less steady state until all of the dark matter in the galactic halo is exhausted, or until the star itself is dynamically ejected from the galaxy. In any event, typical white dwarfs have about twenty cosmological decades (10^{20} years) for life to evolve within their atmospheres. This immense time span is 100 billion times longer than that required for the evolution of life on Earth. Given such a long time, the possibility of biological evolution of some type becomes plausible, and the ascent of complexity is perhaps even likely.

In some respects, the scenario for life in a white dwarf atmosphere is vaguely similar to life on Earth. A white dwarf is about the same radial size as Earth. Just as life on Earth is confined near the planetary surface, any possible life in a white dwarf atmosphere is confined to the outer layers of the star. The inner portion of the star is highly degenerate and chemical reactions in the interior are highly suppressed. Only the outer layer can support interesting chemistry. In the white dwarf, the energy source is a radiation field heating the surface layers from below, whereas Earth is heated from above by light from the Sun. The crucial difference is that life on Earth is based on liquid water, whereas liquid water will be almost nonexistent in a white dwarf atmosphere. In the white dwarf setting, the best one can hope for is that chemical reactions of some type will take place.

The first requirement for life is a proper mix of elements. White dwarfs of higher mass naturally contain large amounts of carbon and oxygen, two of the most important elements in terrestrial organisms. On the other hand, the smallest white dwarfs, with less than half the mass of the Sun, are almost pure helium. Since helium is chemically inert, it is not desirable for an environment with hopes of producing life. Larger white dwarfs thus have a better chance of supporting a biosphere.

For an extended span of time, a white dwarf has a surface temperature of approximately 63 degrees kelvin, close to that of liquid nitrogen. The temperature inside the star is somewhat hotter, although not grossly so. Most of the interior of the white dwarf is highly degenerate, and heat is readily transferred through the star. With this relative ease of heat transfer, the star attains a nearly constant temperature throughout most of its interior. Near the surface, however, the outer layers of the star are made of ordinary matter and hence are not degenerate.

The outermost layer has the capability, in principle, of sustaining chemical reactions and has access to a wide range of photons to drive the chemistry. Inside the stellar core, dark matter annihilation produces high-energy radiation, gamma rays, with energies as large as billions of electron volts. As the radiation works its way to the outside of the star, the radiation is degraded to longer-wavelength, lower-energy photons. At the outer stellar surface, the photons have an average energy of a small fraction of an electron volt. For comparison, chemical reactions have typical energies per particle of a few electron volts. The atmosphere of a white dwarf thus contains the proper range of photon energies to drive chemical reactions.

What about the total energy budget available to such a star? A white dwarf powered by the annihilation of dark matter generates energy at a rate of 10^{15} watts. This power output is small compared to the luminosity of the present-day Sun, but rather large compared to the total power produced by human civilization. As another comparison, the rate at which solar energy is intercepted by Earth is about 10^{17} watts. In other words, the power available to drive biological evolution in a white dwarf atmosphere is 1 percent of the total power accessible to Earth's biosphere.

Let's carry this gedanken experiment even further by making some rough estimates for life forms living in the atmospheres of white dwarfs. Following the lead of Freeman Dyson, we assume that life obeys a kind of scaling law, which in turn implies that the subjective time experienced by a living creature depends on its operating temperature. For lower temperatures, the effective pace of life is slower, and it takes longer for a creature to experience the same number of instants of consciousness.

For our speculative biota evolving near the surface of a white dwarf, the ambient temperature must be near 63 degrees kelvin, about five times less than the temperature of a mammal. The scaling hypothesis says that such a creature requires five times as much real (physical) time to experience the same effective amount of life. Compared to life on Earth, life in white dwarf atmospheres thus loses a factor of five for having a slower metabolic rate and loses another factor of 100 for having less power. This loss of a factor of 500 is more than made up by the time available, which is 100 billion times longer. Combining these two competing effects, we estimate that life in a white dwarf atmosphere has a numerical advantage of about 100 million. Even if the evolution of life in a white dwarf atmosphere is 100 million times less efficient than biological evolution on Earth, the star still has enough time and energy to engender a web of life forms comparable in scope to the biosphere on Earth today.

Our understanding of life and evolution is far from complete. This line of extrapolation does not represent a firm prediction as much as an intriguing statement of possibility. The atmospheres of white dwarfs have a reasonably large energy source and a truly enormous amount of time. In principle, interesting chemistry can take place in this environment. We have no general guarantee that time, energy, and chemistry are sufficient for biological emergence. In the one example we know, however, interesting chemistry has led to the evolution of life. This possibility of the future awaits.

LIFE OUTSIDE WHITE DWARF ATMOSPHERES

A more conventional view of life in the future can be envisioned. White dwarfs powered by the capture and annihilation of dark matter particles provide an effective luminosity of 10^{15} watts. This respectably large amount of power flows out of the stellar surface, which is comparable to the size of the Earth. If a future civilization wanted to use this energy, it could build a spherical shell around the star to intercept its radiative energy. Such an endeavor would require construction on a planetary scale, an expensive but realizable goal for a highly advanced civilization.

For these white dwarf systems, the total power available greatly exceeds the power currently generated and expended by our civilization on Earth. We can put this power rating of white dwarfs into perspective another way. Suppose that the civilization living around a white dwarf has one billion citizens. Each member of the society would then have access to one full megawatt of power, enough to run ten thousand stereos at high volume.

Moreover, this energy supply can last for 20 cosmological decades (100 billion billion years), a great deal longer than the 200-year time span over which we are depleting fossil fuels here on Earth.

THE GROWTH OF BLACK HOLES

Black holes grow larger and more massive throughout the Degenerate Era. They gain mass by accreting stars and gas that come dangerously close to the black hole "surface," the event horizon. As we shall see in the next chapter, black holes must eventually surrender their enormous mass by radiating it away, but not until well after the Degenerate Era has come and gone. In the meantime, they keep gaining weight.

In principle, supermassive black holes could swallow the entire galaxy they live in. How long would this process take? If a one million solar mass black hole, like the one in the center of the Milky Way, accreted stars randomly, it would hoover up the entire galaxy in about 30 cosmological decades (a million trillion trillion years). If the black hole began with a much larger mass, say one billion solar masses, the time required to decimate the galaxy is much shorter, only about 24 cosmological decades. In any case, these times are much longer than the life expectancy of galaxies. As we have already discussed, galaxies will evaporate their stars out into intergalactic space after only about 20 cosmological decades. As a result, most stars escape the immediate wrath of black holes, but some are destroyed in their wake.

Once the galaxies are gone, however, both the black holes and the remaining stellar remnants still exist. After about 20 cosmological decades have passed, the black holes and the remnants belong to their local supercluster, the next larger scale structure that the galaxy once lived within. This larger structure remains gravitationally bound and behaves somewhat like a giant galaxy. The black holes, at least one per former galaxy with cluster membership, will roam freely through the cluster and absorb stars and other matter. The black holes thus continue to gain mass and grow larger.

In the absence of competing physical effects, the dynamical processes of stellar evaporation, gravitational radiation (see Chapter 4), and the accretion of stars by black holes would continue to play themselves out on ever larger sizes and over correspondingly longer times. This hierarchy must come to an end at the close of the Degenerate Era. The stellar remnants, and everything else that we consider as ordinary matter, are composed of protons. And these protons will change their character over enormous spans of time.

PROTON DECAY

One of the surprises coming out of particle physics in the latter part of the 20th century is that the proton will not live forever. Long thought to be stable and infinitely long-lived, protons can decay into smaller particles if given enough time. In effect, protons exhibit an exotic variety of radioactivity. They radiate smaller particles and turn themselves into something new. The time required for this decay process is extraordinarily long, much longer than the current age of the universe, much longer than the lifetimes of stars, and even much longer than the lifetimes of galaxies. In the face of forever, however, the protons will soon be gone.

How can this be? We have already met the positron, the positively charged antimatter partner of the more familiar electron. One might suspect that a proton should decay into a positron, with extra energy left over because the mass of the proton is almost two thousand times larger. The positron thus represents a lower energy state. One of the fundamental governing trends of physics is that systems evolve toward lower-energy states. Water flows downhill. Excited atoms emit light. Light nuclei, like hydrogen, fuse into heavier nuclei, from helium on up to iron, because the larger nuclei have lower energy (per particle). Large nuclei, like uranium, are radioactive and decay to smaller, lower-energy nuclei. Why shouldn't protons decay into positrons or other small particles?

At the most fundamental level, many theories of physics have a built-in law that forbids the decay of protons, even though they could evolve to a lower energy state by doing so. This law can be concisely stated: *baryon number* is conserved. Protons and neutrons are made of ordinary matter that we call baryonic. Each proton or neutron contains one unit of baryon number. Particles like electrons and positrons have zero baryon number, as do photons, the particles of light. So, if a proton decays into a positron, some baryon number is lost in the process.

The newer versions of particle theories, however, contain a loophole. The law that forbids proton decay can be violated, but only occasionally. In practical terms, this apparent oxymoron means that protons will decay over extremely long spans of time, much longer than the current age of the universe.

The proton can decay through many different channels and can thus produce many different decay products. One typical example is shown in the figure below. In this case, the proton decays into a positron and a neutral pion, which will subsequently decay into radiation (photons). Many other decay events are possible. The variety of decay products, and their populations, are not yet known.

One might wonder why we discuss proton decay rather than neutron decay. Inside nuclei, neutrons will decay with a similar lifetime. Free neutrons, however, do not live very long. Left on its own, a neutron will decay into a proton, an electron, and an antineutrino in about ten minutes. This mode of decay is not allowed for neutrons bound into atomic nuclei. Bound neutrons can experience only the long-term decay modes analogous to those of protons.

The average proton lifetime is not well determined by present-day physics. The simplest version of the theory predicts that the time required for proton decay is about 30 cosmological decades (10^{30} years, or a quadrillion quadrillion years). But this simple prediction has already been ruled out by experiments which show that the proton lifetime must be longer than about

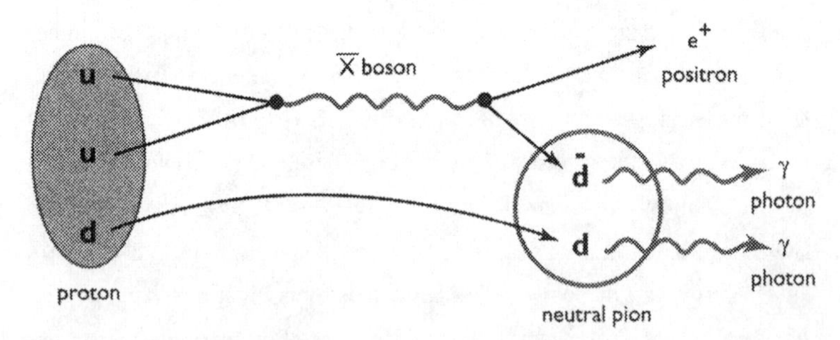

One of the possible channels for proton decay is depicted here. In this case, the end result of the proton decay is a positron (the antiparticle of the electron) and a neutral pion. The pion is highly unstable and quickly decays into radiation (i.e., photons). If the decay occurs in a dense environment like a white dwarf, the positron will quickly annihilate with an electron, producing two more high-energy photons.

32 cosmological decades. The theory that predicts proton decay is a *grand unified theory,* one that involves the unification of the strong, weak, and electromagnetic forces. These theories involve incredibly high energies that have not existed in our universe since the first few instants after the big bang. The energies of the largest particle accelerators are billions of times too small to study this interesting regime of physics. As a result, physicists do not yet have the final version of a grand unified theory. Many possible theories are being studied and they give different predictions for the lifetime of the proton.

Given that the universe is only ten billion years old, the idea of performing an experiment to measure a lifetime of a quadrillion quadrillion years (30 cosmological decades) seems virtually impossible. The basic idea is quite simple, however, once the general concept of radioactive decay is understood. All of the particles, in this case protons, do not live for a fixed amount of time and then all decay at once. Instead, particles have a chance to decay *at any time.* Since it is unlikely for a decay to take place, the odds are highly against it, most particles will live to a ripe old age. The particle lifetime is an *average* time that particles live, but not the *actual* life span of every particle. There will always be some particles that decay early. And this particle version of infant mortality can be measured experimentally.

To detect the decay process, one needs a large number of particles. To fix ideas, suppose we want to measure the decay of a proton with an expected lifetime of 10^{32} years. If we get a large box containing 10^{32} protons (a small swimming pool 20 meters long, 5 meters wide, and 2 meters deep would do the trick), then about one proton per year would decay within the experimental apparatus. If we could build sensitive instruments to detect each and every decay event, then all we would have to do is wait a few years and our measurement would be done. In practice, the experimental difficulties associated with these measurements are a little more subtle, but the basic idea is clear. In particular, we don't need to wait 10^{32} years to find out the answer. Experiments of this type have already shown that the proton lifetime is longer than 10^{32} years. And more experiments are now being conducted to continue the search for proton decay.

Proton decay can be predicted in very general terms. In the early universe, some process involving baryon number violation produced the matter that we see in the universe today. Recall that a slight excess of matter over antimatter was generated during the first microsecond of cosmic history. The universe can have more matter than antimatter only if some physical process generates extra baryon number. But if such a baryon number violating process can occur, the protons are doomed. It is only a matter of time before they must decay.

The proton decay channels alluded to thus far do not involve the fourth force of nature—gravity. An additional mechanism for proton decay is driven by the gravitational force. The proton is not really a single particle, but rather is composed of three constituent particles known as quarks. The quarks inside a proton do not sit around at rest—they live in a state of constant agitation. Every once in a great while, the quarks will occupy nearly the same position inside the proton. When this convergence occurs, if the quarks are close enough together, they can coalesce into a microscopic black hole. The estimates for the average time required for a proton to tunnel into a miniature black hole vary widely, anywhere from 45 to 169 cosmological decades with some preference for the lower end of this range. Needless to say, this process is poorly understood and the corresponding lifetime is recklessly uncertain. But, unless they decay sooner, protons are destined to disappear through this process—death by the force of gravity.

As we discuss in the next chapter, black holes do not live forever. And small black holes live for much less time than larger ones. After a proton transforms itself into a black hole, it will evaporate almost instantly and leave behind a positron. The proton thus provides yet another theater for the war between gravity and thermodynamics. Because of the relentless nature of gravity, it eventually can instigate the demise of protons and the formation of tiny black holes. But this apparent gravitational victory is short-lived. The black holes evaporate immediately after they are made. Most of the mass-energy of the proton is radiated away, entropy is released into the universe, and thermodynamics carries the day.

Protons can decay through yet another, even more exotic, mechanism. The vacuum configurations of empty space can have more than one possible state. The vacuum can in principle spontaneously change its configuration through a process of quantum mechanical tunneling. Since transitions between the different vacuum states produce a change in baryon number, these vacuum transitions will drive proton decay. Such transitions are highly suppressed, however, and take an extraordinarily long time. In the absence of a faster decay channel, protons will be destroyed by this mechanism during cosmological decades 140 to 150.

THE ULTIMATE FATE OF DEGENERATE REMNANTS

The final chapter of stellar evolution plays itself out as the protons decay. Although the true proton lifetime has not been measured experimentally, we adopt a representative proton lifetime of 37 cosmological decades (ten trillion trillion trillion years) throughout this discussion. When protons decay inside a star, a white dwarf for example, energy is added to the stellar budget. The most common decay products are a positron and a pion, which immediately decays into high-energy gamma radiation. The positron quickly finds an electron and the two annihilate into two more high-energy gamma ray photons. The net result is thus to convert the *rest mass energy* of the proton into gamma radiation, which heats up the star. Decaying protons thus endow the star with an internal energy source, but only at a high price: the star must surrender its own rest mass energy to create the heat and light.

A white dwarf powered by proton decay has a luminosity of about 400 watts, barely enough to run a few light bulbs. An entire galaxy of such stars has a luminosity ten trillion times smaller than the Sun. Even if we add together the power from all the stars in all the galaxies currently within our cosmological horizon, the total would still be one hundred times less than that of the Sun. This future is dim indeed.

Inside the white dwarf, radiation must scatter many times before it can diffuse its way out to the stellar surface. During this future epoch, the surface temperature of a white dwarf will be a frigid 0.06 degrees kelvin, about 100,000 times colder than the Sun. As a result, these 400-watt light bulbs do not make very good reading lamps. They emit radiation with a characteristic wavelength of 5 centimeters, about 50,000 times too long for human eyes to see.

During the evolutionary phase when protons decay, the chemical makeup of a white dwarf changes completely. Suppose we begin with a star made of pure carbon. Each carbon nucleus contains six protons and six neutrons. As the protons and neutrons decay, the nuclei become smaller and contain fewer particles. As this process continues, the original carbon nuclei are reduced to a single particle and the star ends its life as pure hydrogen.

Two effects complicate this simple picture. The first complication is that high energy radiation resulting from proton decay can free other protons and neutrons from their homes within nuclei. These emancipated particles generally give up their newly found freedom and join other nuclei. On average, each proton decay event leads to one additional proton or neutron moving from one nucleus to another. We thus obtain a type of nuclear musical chairs.

The second complication is cold fusion. Even at low temperatures, in this case less than one degree above absolute zero, nuclei can sometimes fuse because of the Heisenberg uncertainty principle. The wavelike nature of the particles makes it impossible to pinpoint their precise location. As a result, two nuclei occasionally find themselves close enough to fuse into a heavier nucleus. Inside a white dwarf, which is a million times denser than Earth, cold fusion takes only one hundred thousand years for hydrogen and about 200 cosmological decades (10^{200} years) for carbon. White dwarfs thus tend to retain a helium composition. These time scales are long enough, however, that cold fusion does not greatly alter the evolution of a white dwarf during its proton decay phase 10^{37} years into the future. One can also easily understand why cold fusion does not play an interesting role in the universe of today.

As a white dwarf continues to lose mass through proton decay, its structure changes markedly. Because of the counterintuitive nature of degenerate matter, a white dwarf expands in radial size as it withers away. As the star expands, the density becomes lower, and the matter eventually ceases to be degenerate. This transition occurs when the stellar mass falls to that of Jupiter, about one thousand times smaller than the mass of the Sun. At this point in its evolution, the star has the density of water and a radius ten times smaller than the Sun. The star consists of a frozen array of hydrogen atoms—a large ball of hydrogen ice.

After degeneracy has lifted, the crystalline white dwarf continues to waste away until it is so small that it can no longer function as a star. This final transition marks the end of stellar evolution. True stellar death occurs as the object becomes transparent, when radiation propagating inside the star can escape freely with no diffusion. At this turning point, the star contains only 10^{24} grams, about six thousand times less mass than Earth's.

The penultimate fate of most stars is thus to end up as a hydrogen rock, about seventy times smaller than the Moon. The rock continues to evaporate away its mass as proton decay sputters toward completion. The ultimate fate of white dwarfs thus becomes clear: nothing is left. The entire mass energy of the star is eventually radiated into interstellar space. Once again, thermodynamics triumphs over gravity in the long run.

The neutron stars, those rare and dense relatives of the white dwarfs, evaporate in a similar fashion. Proton decay empowers the neutron stars with approximately the same total luminosity, about 400 watts. Neutron stars are much smaller than white dwarfs. In order to radiate the same total power, the stellar surfaces must be hotter, about 3 degrees kelvin for a typical neutron star. This temperature is about the same as the present-day cosmic background radiation, which sets the lowest temperature available in the universe of today. In contrast, during the 37th through the 39th cosmological decades, neutron stars softly glowing at 3 degrees kelvin will be among the hottest objects in the universe.

The final phases in the life of a neutron star play out somewhat differently from the white dwarfs. As a neutron star loses its mass through proton decay, it grows less dense and neutron degeneracy is eventually lifted. When the neutrons are no longer degenerate, they transform themselves into protons, electrons, and antineutrinos. This transition occurs when the mass of the star drops below one tenth of a solar mass and the stellar radius is about 164 kilometers. At this point, the density is still large enough that electrons remain degenerate, and the star closely resembles a white dwarf. The remaining white-dwarflike entity continues to lose mass as more protons decay, until the electron degeneracy is lifted also. The object then becomes a rock of hydrogen ice, with less than one-thousandth of a solar mass. The protons in the crystal lattice also decay, eventually evaporating the entire star into radiation and small particles. The ultimate fate of neutron stars is for nothing to remain.

The long-term fate of the planets tells a similar story. Planets are also made primarily of protons, which decay as the planet evaporates into radiation. By the time proton decay begins to compromise their structure, the remaining planets have long since been wrenched away from their parent stars and are drifting in isolation. As they slowly sublimate, planets generate a rather modest power output, only about one milliwatt for an Earthlike planet. Although planets begin with a larger admixture of heavy elements than the stars, they too are reduced to hydrogen ice in due time. Even a planet composed of pure iron is broken down by the 38th cosmological decade, about six proton half-lives. During the 39th cosmological decade, the planet evolves from a small lump of hydrogen crystal into a completely disintegrated state.

By the 40th cosmological decade, nearly all of the protons in the universe have decayed and the degenerate stellar remnants are gone. The seemingly resolute and indestructible stellar remnants have given way to a diffuse sea of radiation, mostly photons and neutrinos, with a small admixture of positrons and electrons. The universe takes on a new character. This vast arena of overwhelming desolation is punctuated by isolated regions of extreme space-time curvature, in other words, black holes. As the Degenerate Era draws to a close, the black holes, containing from one up to billions of solar masses, push doggedly forward into the next epoch.

4

THE BLACK HOLE ERA

$40 < \eta < 100$

Black holes inherit the universe, warp space and time, evaporate their mass energy, and make an explosive exit.

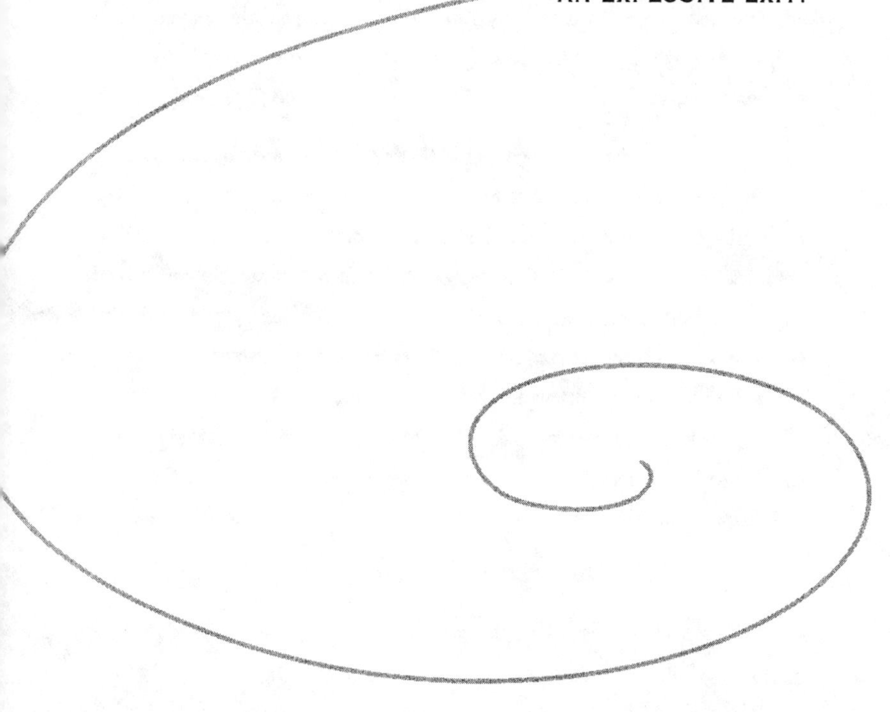

Cosmological decade 90, on the outskirts of a black hole cluster:

Bob was confused. His diffuse body, containing more than a million solar masses, was awash in a frightening kaleidoscope of sensations. He had never experienced anything remotely like this mental electrical storm, despite being nearly 10^{79} years old. The wave of distortions reached an almost unbearable peak before suddenly dying away, leaving only a residual queasiness, a feeling that might be described as seasickness if the concepts of water and oceans were available. The positrons and electrons of his brain traced a slow spiral of calculations, and he gradually realized that a burst of gravitational radiation had caused the discomfort. Somewhere in the distance, two black holes had coalesced, carving out a gravitational potential well of stupendous depth. Black holes were one of the few remaining dangers to members of his species, so Bob was relieved that this particular celestial collision had taken place a long way off.

Of course, Bob had been lectured on the ultimate importance of black holes. Their eerie glow bathed the universe in a rarefied sea of radiation and afforded the energy source that allowed virtually everything, even life itself, to exist. He knew that without the essential energy provided by black hole evaporation, the universe would be a lifeless and boring place.

Although Bob's concerns tended to be of a flatly practical nature, there were members of his species who were trying to understand the properties of the universe during its first 10^{40} years, "those almost unimaginably brief moments after the big bang." Particularly fashionable was a wild conjecture that highly complex structures might have been based on the interaction of electrons with protons and neutrons. The existence of protons and neutrons, exotic short-lived particles that had long since decayed away, was enthusiastically embraced by the more adventurous physicists of the time, and simultaneously denounced as "rampant speculation" by those cut from a more conservative cloth.

As protons decay, the universe loses the dust, the white dwarfs, the frozen Earth, and the material of everyday existence. After the protons are gone, the fabric of the universe is dramatically altered. The most significant remaining objects are black holes, which slide unscathed through the close of the Degenerate Era. Black holes are stellarlike objects, albeit with some very unusual properties, and upon outlasting the white dwarfs, they inherit the role now filled by ordinary stars. When the universe reaches its 40th cosmological decade, the black holes reign supreme. They provide the light, the heat, and the dynamics to keep things interesting.

The black holes are scattered through an inordinately rarefied sea of elementary particles. Imagine sifting through volumes of this nearly perfect vacuum. One encounters an occasional electron, the negatively charged particle that orbits the nuclei of today's atoms and flows through wires in electrical circuits. A persistent search reveals that every surviving electron has a corresponding antimatter partner, a positron. Each positron carries a single positive charge, and so the universe as a whole always remains electrically neutral. Further searching turns up more stealthful inhabitants of the interstellar gulfs—axions, different flavors of neutrinos, and more.

The universe of the Black Hole Era is steeped in low energy photons, light with wavelengths far too long for human eyes to see. Light that our eyes can detect is composed of photons with wavelengths of about one half of a micron (one half of one thousandth of a millimeter). The wavelength of typical radiation in the 40th cosmological decade is much longer, nearly a kilometer. In order to "see" in the Black Hole Era, one needs eyes the size of continents.

BLACK HOLES DEFINED

What is a black hole? The conventional definition might run: A black hole is an object that distorts the space-time continuum so severely that even light cannot escape from the object's surface. This chapter will explore the meaning of this statement in detail, but the basic idea behind a black hole's phenomenal gravity is readily grasped.

Nearly everyone has seen grainy footage from the Apollo Moon landings. The astronauts bounce around easily in their bulky spacesuits. Their stiff-legged jumps exhibit an unearthly quality for good reason: on the Moon, the force of gravity is six times weaker than on Earth. A ball thrown upwards with a given speed goes higher above the Moon than it goes above Earth. Likewise, less energy is needed to escape from the Moon's gravitational clutches.

The speed required to completely overcome the gravitational pull of a body is called the *escape speed.* To escape Earth, for example, a velocity of 25,000 miles per hour (11 kilometers per second) is required (in the absence of air friction). Enormous Saturn V boosters supplied this necessary speed for the Moon shots. In contrast, the comparatively modest rockets on the lunar modules sufficed to leave the Moon and return to Earth. The Moon's escape velocity is small because the Moon is both less dense and less massive than Earth. If two objects have the same size but have different masses, the escape speed is greater for the more massive object. For example, if we imagine an object with the mass of the Sun and the diameter of Earth, the escape speed from this dense altered world would be about 6500 kilometers per second, 588 times larger than Earth's true escape speed. If we increase the mass while keeping the diameter fixed, the escape velocity increases, and it gets much harder to blast away from the surface. Eventually, after more than two thousand solar masses are squeezed inside the Earth-sized sphere, the escape velocity exceeds the speed of light (300,000 kilometers per second). If the escape speed is greater than the speed of light, then nothing, not even light, can escape. Our dense imaginary sphere becomes a black hole. The name is eminently appropriate: an object that emits no light appears black to the external universe.

Although nothing travels fast enough to escape from a black hole surface, a black hole is *not* a cosmic maw destined to consume everything in its vicinity. The gravitational pull of any object, black holes included, decreases as one moves away. At a sufficient distance, a black hole's gravity is indistinguishable from the gravity of an ordinary star of comparable mass. Far from the black hole, the local escape speed is always less than the speed of light, and particles or spaceships can come and go at their leisure. As one draws close to a black hole, the escape speed steadily increases. At a well-defined radius, the escape speed finally exceeds the speed of light. This point of no return marks the location of an effective surface for the black hole, named the *Schwarzschild radius* in honor of Karl Schwarzschild, a German physicist who was among the first to absorb Einstein's theory of general relativity. Shortly after deriving the radius that now bears his name, Schwarzschild died from a disease contracted at the Russian front during World War I.

The Schwarzschild radius of a one-solar-mass black hole is a few kilometers. One can visualize the close-packed nature of such a body by imagining the entire Sun compressed to the size of a small college town. The Schwarzschild radius increases in strict proportion to the mass of a black hole. A million-solar-mass black hole, for instance, has a radius of three million kilometers, about four times larger than the present-day Sun. If Earth were compacted down to form a black hole, it would be about the size of a marble. Imagine the Sears Tower, Mount Everest, Earth's vast iron core, and all of the minimalls known to mankind crammed into a sphere that fits easily in the palm of your hand! Although astonishing, such a bizarre object can actually exist.

The spherical surface marked by the Schwarzschild radius encloses a portion of the space-time continuum so severely warped that no particles can escape. Since no information can travel out of this local region, it is effectively separated from the rest of the universe. We can thus think of the Schwarzschild radius as defining an imaginary surface that provides the boundary between the rest of the universe and this interior region from which no information can escape. This boundary is called the *event horizon.*

Black holes are thus a bit like the vault in a bank. Information locked up inside the event horizon is spirited away from the rest of the universe. However, the bank vault has a key and someone can remove the information, or perhaps the cash, stored within. For a black hole, no one has a key. The information is locked away forever . . . or rather, almost forever.

TYPES OF BLACK HOLES

Viewed one way, black holes come in several different varieties. Viewed from another perspective, however, black holes are all essentially the same. These different points of view depend on historical considerations.

If we consider their past histories—the processes that form black holes—we find that several different mechanisms produce black holes and that each mechanism makes black holes within a particular mass range. Stellar mass black holes are produced by massive dying stars, whereas the largest black holes are manufactured by galaxies. Much smaller black holes might result from exotic processes that transpired during the tiny fragments of time immediately following the big bang. When the Black Hole Era finally arrives, at least two different varieties of these ghostly objects assume their leading roles.

Independent of the production mechanism, once a particular black hole has formed, its characteristics are completely described by only three quantities: the mass, electric charge, and rotation rate. Except for these three defining quantities, nothing remains of a black hole's history. Because the past is effectively swallowed up, black holes are very homogeneous objects and are devoid of intractable individual complications. This stark simplicity is often summed up with the pronouncement that "a black hole has no hair."

The most common type of black hole is manufactured in the course of stellar evolution. When sufficiently massive stars end their lives by exploding in supernovae, they sometimes leave behind a stellar remnant that is too massive to be supported by degeneracy pressure. The dense remnant is gravitationally unstable and immediately collapses to become a black hole. These *stellar black holes* have masses comparable to ordinary stars, although they are about one hundred thousand times smaller in size. For example, a relatively nearby black hole candidate, Cygnus X-1, has a mass about ten times that of the Sun, and a radius of roughly 30 kilometers.

Such stellar black holes seem insignificant compared to a *supermassive black hole* which anchors the core of a galaxy. Black holes in the supermassive category can encompass billions of solar masses, and have Schwarzschild radii larger than the size of Jupiter's orbit around the Sun. Astronomers now think that virtually every galaxy has a giant black hole residing at its center. The center of our own galaxy appears to harbor a rather modest supermassive specimen; recent measurements suggest that it weighs as much as three million Suns.

Although the origin of galactic supermassive black holes is not completely understood, they arose when primordial clouds of gas collapsed to form galaxies. After a galactic collapse gets underway, so much material, newborn stars and gas, collects in the nascent galaxy's center that a black hole is compelled to form. Once a black hole is born, the crowded environs of the young galactic core naturally encourage the hole to consume additional material that strays too close, and the black hole steadily grows larger. Quasars, the extremely luminous cores of distant and ancient galaxies, are powered by supermassive black holes that gather dust, gas, and torn-up stars into a whirlpool-like accretion disk. As gas funnels into the central black hole, friction heats up the disk and powers a brightly shining quasar.

Our third category, the *primordial black holes,* rests on a more speculative footing than the first two varieties. If primordial black holes exist, they must have been created shortly after the big bang. They are also likely to be very small, much less massive than stars. Like a diamond forged in a kimberlite pipe, a tiny primordial black hole is a relic of monstrous pressures and temperatures, conditions that held sway early in cosmic history.

In spite of their inviolable reputation, black holes will not last forever. As we shall see, the energy produced by the slow evaporation of black holes powers the universe during the Black Hole Era. Larger, more massive black holes live much longer than smaller ones. In fact, the smallest black holes with masses less than 10^{25} grams, which presumably have a primordial origin, evaporate before the Black Hole Era even starts. The maximum size of these early-expiring black holes is only about one hundredth of a millimeter, although they weigh as much as a respectably large moon in our solar system. Larger black holes with masses greater than one six-hundredth of Earth's mass will survive to participate in the Black Hole Era.

BASIC BLACK HOLE WEIRDNESS

Black holes make a mockery of our common sense ideas concerning space and time. During the Black Hole Era, the universe becomes a very alien place. From our current vantage point in the Stelliferous Era, black holes are a mere curiosity. They have little bearing on our Earthbound lives and do not significantly affect the evolution of the galaxy. In the Black Hole Era, however, general relativity and black holes reign supreme—they are absolutely essential to the everyday business of running the universe. Before pushing the clock beyond the 40th cosmological decade, we need to examine several ideas from the theory of general relativity. We can then taste the intriguing blend of simplicity, elegance, and sheer strangeness that characterize black holes.

Imagine a rocket floating through space, far away from the gravitational influence of any planets or stars. The engines are cold and the rocket drifts in a straight line at a constant speed. Inside the hold, the cargo is weightless. Astronauts hang suspended, and Tang transforms itself into glistening orange spheres as the rocket and its contents travel through space together. Suddenly, someone turns the rocket engines on at full power. Exhaust roars away from the thrusters, the hull pushes forward, and the startled astronauts find the floor rushing up to meet them. The Tang splatters. As the engines continue to fire, the rocket accelerates and its speed continually increases. The astronauts stay pressed onto the floor.

In a rocket equipped with weak engines, the acceleration is not very rapid. If the astronauts leap toward the ceiling, the rocket floor would only gradually catch up with them. They would feel like they were walking on the Moon, where gravity is weak and jumping is easy. As the engines produce larger accelerations, however, the floor of the rocket comes to mimic the surface of Earth. That is, if the engines are adjusted so that during each second the rocket's speed increases by 9.8 meters per second, then objects fall to the floor in exactly the same way that they fall to Earth's surface. Most importantly, if the rocket has no windows, the space travelers inside have no way of knowing whether the acceleration they feel is due to the force from the rocket engines, or due to the gravity of a planet.

At first glance, this statement of equivalence, which equates the effects of gravity and acceleration, might not seem particularly intriguing. Einstein realized, however, that the equivalence between gravity and acceleration is extraordinarily profound. He used this concept as a point of departure in developing the theory of general relativity. In turn, general relativity has enlightened us about the strange properties of black holes.

An unswerving acceptance of the equivalence principle forces us to abandon our everyday conception of absolute time. Consider a clock taped to the floor of a large rocket, and another clock fixed to the ceiling. The clocks are designed so that they broadcast a radio announcement of the time: "The time at the tone is . . ." When the rocket is coasting through empty space, an astronaut floating just above the floor of the rocket can check the time being kept by the clocks by listening to the radio. At the turn of an hour, the clocks broadcast a tone. The astronaut receives the signal from the clock next to him on the floor, and then, a tiny fraction of a second later, he picks up the tone from the ceiling clock. The exceedingly slim, but nevertheless real, difference in arrival times between the two tones arises because the ceiling clock is further away. The ceiling clock's radio signal, traveling at the speed of light, takes slightly longer to reach the astronaut than the signal emitted by the nearby floor clock. The astronaut concludes, after carefully comparing several hours' worth of clock signals, that clocks are marking time at the same rate, but that the ceiling clock is set slightly behind (by the ceiling-to-floor light travel time). All clocks on the rocket run at the same rate if the rocket is coasting through space at a constant velocity.

When the rocket accelerates, the situation becomes more interesting. An astronaut standing on the floor observes no change in the rate of the clock that stands next to him. Because of the upward acceleration of the rocket, however, he observes that the ceiling clock is running faster. The ceiling clock sends a tone (in the form of a radio wave) down to the floor. Because the floor is accelerating upwards, it intercepts the radio wave sooner than if the rocket were merely coasting along. If the acceleration continues, subsequent tones also arrive earlier than expected. In the viewpoint of the astronaut on the floor, the ceiling clock is broadcasting its time intervals at an increased rate, and is running fast compared to the floor clock.

According to the equivalence principle, the phenomenon of mismatched clock rates, which occurs in response to the acceleration of a rocket, *also* occurs in a uniform gravitational field. The equivalence principle therefore insists on a seemingly bizarre conclusion. Two clocks at different heights above Earth's surface must measure the flow of time at different rates. This strange behavior is an intrinsic feature of gravity. The variation of the flow of time within a gravitational field is entirely independent of the mechanism used to measure time. Atomic clocks, quartz watches, and biological rhythms all experience the passage of time to be dilated or compressed in the same manner.

We have no intuitive sense of relativistic time dilation because its effects are completely negligible in our environment. In everyday life, the malleability of time in response to gravity is ludicrously small. For instance, suppose that identical twins spend the night in bunk beds. One sleeps a meter above the other. The next morning, the twin who slept in the top bunk is a few trillionths of a second older. Although the twins won't live long enough for gravitational time dilation to become a major source of sibling rivalry, Earth itself will live for 38 cosmological decades. Over this long span of time, this time dilation creates a time differential of 10^{22} years, or 22 cosmological decades.

The strong gravitational field of a black hole leads to extremely severe time dilation. Close to a black hole, clocks run far slower than in empty space. Near the event horizon, the effective surface of the black hole, clocks come almost to a complete stop. As with all effects due to relativity, the exact meaning of this statement depends on the observer. A stationary observer hovering near the event horizon is pressed against the floor of his spaceship, and experiences the crushing strength of the black hole's gravity, as well as a profoundly slowed passage of time. A freely falling observer, however, has no impression of the gravitational force, feels no acceleration, and experiences time at the usual rate.

Black-hole-induced time dilation gives rise to some remarkable possibilities. Clocks put into special orbits near a black hole run exceedingly slowly. In fact, with a careful choice of orbit, time can be slowed down by an arbitrarily large amount. This effect could be exploited by inserting protons, for example, into the special orbit. The protons would live in an environment where the flow of time has essentially ground to a halt. They would remain in existence long after ordinary matter has decayed in the external universe, where time runs at the normal rate. Long-term storage does not come for free, however. If someone in the future wants to withdraw the stash of protons, say to build a machine or to grow a clone, the particles would have to be extracted from their orbits at a considerable expense of energy. The requirement would be about the same as that needed to create protons out of pure energy in accordance with Einstein's well-known formula, $E = mc^2$.

Another far-out possibility is to place an observer, either a person in a spaceship or a robotic machine, near a black hole in an orbit subject to extreme time dilation. The observer could then effectively "travel" into the future. In an extreme case, for example, the time dilation could be so severe that the time traveler would experience only one year of time while the outside universe (far from the black hole) records 60 cosmological decades. This effect gives adventurers the possibility of a journey into the far future, to the slow-paced reaches of the Black Hole Era.

CURVATURE

Black holes distort the structure of space just as severely as they warp the passage of time. The space (or more precisely, the space-time) in the vicinity of a black hole is strongly curved. We have already met the concept of curved space in discussing the expansion of the universe as a whole. If the universe is open, the geometry of space over the largest scales has a *negative* curvature. If the universe is closed, then space has a *positive* curvature. If the universe lies exactly on the dividing line between being open and closed, then space is *flat* (see the figure on the next page).

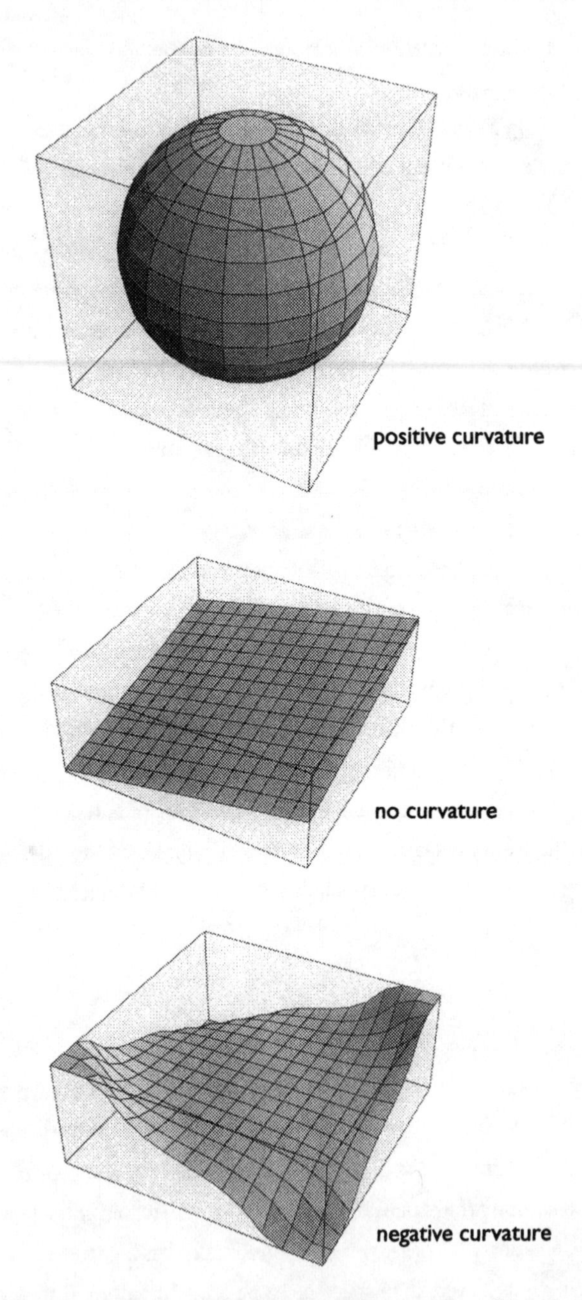

The gridded surface in the bottom image is shaped like a saddle, and is said to exhibit a negative curvature. The surface in the middle image is flat, and in turn is said to exhibit zero curvature. The sphere in the top image is positively curved.

What does it mean for the geometry of space to be curved? The curvature of three-dimensional space is extraordinarily difficult to visualize. When we think of space, our intuitive view is a flat Euclidean expanse, extending uniformly in three perpendicular directions. The concept of space curvature is completely alien because the amount of curvature intrinsic to the space in which we live is extremely small. Our ability to visualize is borne of evolutionary pressures, and because the space of our world is almost perfectly flat, no evolutionary advantage could be gained from the ability to visualize a curved three-dimensional space. Only in the past hundred years have we faced the need to think seriously about curved space and its attendant non-Euclidean geometry.

Although curvature in three dimensions is a challenging concept, the idea of curvature in a two-dimensional space is easy to grasp. Let's reexamine our everyday notion of a circle. In two dimensions, a circle is the collection of points which lie at a particular distance from a central point. In an ordinary flat two-dimensional plane, the distance around a circle (the circumference) is $\pi = 3.14159\ldots$ (*pi*) times longer than the distance across the circle (the diameter). This relationship can be verified by making careful measurements of a circle drawn on a piece of paper.

Next, suppose that we could measure distances around large circles drawn on the surface of Earth. To carry out this experiment, we find a perfectly smooth plain (for instance, the Antarctic polar plateau) and then tie a long string to a fixed post (located perhaps at the South Pole). We then measure the distance covered in walking in a circle around the pole. If we make several measurements of this sort, each time using a longer string, we discover something curious about the resulting circles. With a ten-mile-long string, the diameter of the circle is 1.000001 times too long, that is, longer than the diameter we expect from dividing the measured circumference by π. With a 100-mile-long string, the diameter of the circle is 1.0001 times too long. If we make the string 6250 miles long, so that it stretches from the South Pole to the Equator, the distance across the "circle" would be one-half of the distance around the circle, 1.57 times too long.

Circles marked out on the ground behave this way because Earth has a curved two-dimensional surface. We can easily visualize the curvature of a spherical surface because we can see how it is embedded in flat three-dimensional space. Unfortunately, however, the visualization of a curved three-dimensional space requires us to step up to another dimension and visualize the embedding four-dimensional space. Such a feat is very difficult for the human mind to perform.

If a curved surface behaves like the surface of Earth, in the sense that a circle's diameter is longer than its circumference divided by π, then we say that the surface has a positive curvature. Likewise, if the diameter of a circle is less than its circumference divided by π, the curvature is negative.

The theory of general relativity tells us that mass induces a curvature in three-dimensional space. The radius of an imaginary sphere which surrounds a massive body is slightly longer than what one would infer from a measurement of the distance around the sphere's equator. Equivalently, the actual volume contained within a sphere that surrounds a concentration of mass is greater than the volume that one would predict from measuring the surface of the sphere and then applying formulas from ordinary Euclidean geometry.

The curvature produced by Earth's mass is very small. The distance to the center of the Earth is about 1.5 millimeters farther than the circumference of the planet divided by 2π. In this sense, the geometry of our local space departs from perfect flatness by one part in four billion. If Earth had more mass, it would produce more curvature. Objects with greater mass can produce correspondingly larger curvature. The distance from the center to the surface of the Sun, for instance, is about a half kilometer longer than the circumference of the Sun divided by 2π. White dwarfs and neutron stars, which are far denser than the Sun, produce much larger curvatures in their immediate vicinities. The distance to the center of a neutron star is almost 10 percent greater than the stellar circumference divided by 2π.

The curvature of three-dimensional space near a compact star can be illustrated with a graphic device called an *embedding diagram*. To make one of these diagrams, we first imagine cutting a star in half. We then show the intrinsic curvature in the equatorial plane by plotting that plane as a curved two-dimensional surface in a flat three-dimensional space (see the figure opposite). Using this trick, we can get an idea, or representation, of the spatial relationships between points that lie on the equatorial cross-section of the star. The embedding diagram shows that the slice of space which passes through the neutron star is positively curved. In the diagram, the graphical downwarping of the two-dimensional equatorial plane shows why the diameter of a compact star is surprisingly large compared to the circumference. In other words, the space occupied by the star depicted by the embedding diagram is positively curved.

Near a black hole, the local curvature of space is so strong that the distance to the center of the black hole is infinitely longer than the distance around the circumference (the Schwarzschild surface of the hole). This infi-

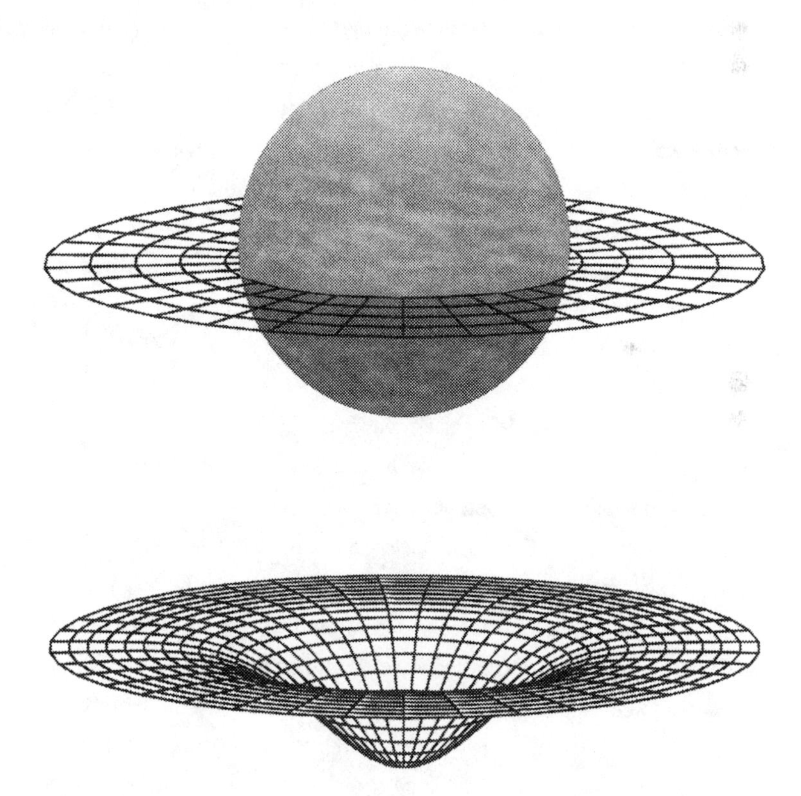

An embedding diagram which helps in the visualization of the curvature of space produced by a massive spherical body. The downwarped surface indicates the spatial relationship between points lying on the cross-sectional cut through the dense sphere.

nite deformation of space inside a black hole is what makes these objects seem so strange. The curvature creates the event horizon, the curvature causes the extreme differences in the passage of time for different observers, and the curvature bestows on large black holes the fortitude to last into the extremely distant future.

TIDAL FORCES

Einstein showed that the curvature of space caused by a concentration of mass produces a *tidal force*. Stellar black holes derive a great deal of their infamy from the extreme nature of the tidal forces near their event horizons. But what exactly is a tidal force? The answer is best given by way of an example. If you stand on Earth's surface, your head is slightly farther away from the center of Earth than your feet. Since the gravitational force exerted on you by Earth gets weaker with increasing distance, the gravitational force on your head is marginally less than the force on your feet. Because of this difference, your feet are being pulled away from your head. Fortunately, this tidal effect is rather small. Because the curvature of space produced by Earth is extremely tiny, the tidal stretching force is about two million times smaller than the force of gravity itself, and we don't notice this extra force as we walk around on Earth's surface.

Near the surface of a stellar black hole, however, the tidal forces are enormous. To fix ideas, imagine being near a black hole with the mass of the Sun. The Schwarzschild radius is only about 3 kilometers. If you could stand on the black hole surface, your feet would once again be pulled away from your head through tidal stretching. This time, however, the force is one billion times larger than the gravitational force you feel from Earth. In other words, your feet will be pulled away from your head with about one hundred billion pounds of force. Any ordinary macroscopic object—a rock, a space probe, an astronaut—would be completely ripped apart by these powerful tidal forces.

Tidal forces produce an additional effect. They squeeze objects together in the perpendicular direction. Since gravitational forces are always directed radially inwards toward the center of the black hole (or any massive object), the force acting on one side of a body is not quite in the same direction as that on the other side. This difference in direction gives rise to a squeezing effect. As in the case of tidal stretching, the magnitude of this extra force is extremely small in a weak gravitational field, such as that on Earth's surface. Close to a black hole, however, these squeezing forces can be enormous, roughly the same strength as the tidal stretching forces. Imagine standing near the surface of a black hole with the mass of the Sun. While your head is being pulled away from your feet by one hundred billion pounds of force, your sides would be crushed together by this same amount of force—a rather unpleasant prospect.

If a black hole gains mass, its gravitational influence increases, but the tidal forces near the surface become less severe. In the vicinity of a billion-solar-mass black hole, for example, the tidal forces are modest enough to allow an astronaut to cross the event horizon without suffering great discomfort. However, these supermassive black holes pose their own particular hazard to future space travelers. The weak tidal effects and the total blackness of such a black hole make it possible to cross the Schwarzschild radius—the point of no return—before it becomes apparent that anything is amiss.

A MENAGERIE OF BLACK HOLES

When the Black Hole Era finally arrives, how many black holes will the universe contain? Let's make a ballpark estimate. Nearly every present-day galaxy, including our own, has a supermassive black hole anchoring its center. These enormous central black holes weigh in between one million and several billion solar masses. We also know that additional supermassive black holes are not roaming *en masse* through the outer galactic reaches. If large populations of supermassive holes were orbiting within the halos of spiral galaxies, they would wreak havoc upon the delicate galactic disks by severely disrupting the orbits of stars. The absence of this destructive process suggests a simple accounting scheme for the number of supermassive black holes—one per galaxy.

The stellar black holes produced by supernova explosions constitute a more numerous population than the supermassive black holes. About three stars out of every thousand are born with enough mass to develop an iron core and explode as a supernova. The percentage of supernovae that leave behind black holes, as opposed to neutron stars or complete dispersal, lies in the neighborhood of 1 to 10 percent. Each galaxy will thus contain roughly one million stellar black holes by the time star formation and stellar evolution have run their course. These stellar black holes have masses ranging from about three to one hundred solar masses, with the average black hole falling near the low end of this range.

By the 40th cosmological decade, the galaxies are completely destroyed. Galactic gas and dust will have long since been dispersed or incorporated into stars. Because of the decay of their constituent protons, stellar remnants, brown dwarfs, and planets have entirely evaporated. Except for stray radiation and scattered elementary particles, the sole legacy of each galaxy is one supermassive black hole and about one million stellar black holes. As these black holes slowly trace their orbits through vast gravitationally bound clusters, some fraction of the supermassive black holes merge to form gargantuan conglomerate black holes. The volume of space which now makes up our observable universe contributes roughly thirty billion supermassive holes and thirty million billion stellar holes to the overall melange. The total number of black holes in the entire observable universe at the start of the Black Hole Era is colossal. The horizon is 10^{30} times farther away than it is now. If the large-scale geometry of space-time is flat, the observable universe will contain about 10^{40} black holes with supermassive status, and nearly 10^{46} stellar black holes, one trillion trillion holes for every star within the cosmological horizon of today's universe.

BLACK APOCALYPSE

During the Stelliferous and Degenerate Eras, black holes steadily grow larger and more massive as they accrete material from the surrounding universe. The accretion of material occurs through the action of ordinary gravitational forces—black holes do not operate as indiscriminate cosmic vacuum cleaners.

For objects substantially beyond the Schwarzschild radius, the black hole acts just like a star or any other astronomical object with a large mass. If a space probe drifts near a black hole, for example, the hole's gravity will generally whip the probe through a hyperbolic orbit and send it flying away in a new direction. The probe can cross the event horizon only if it is aimed directly at the black hole center. In practice, most accretion onto black holes occurs when material, gas or dust, collects in a circular orbit around the black hole by forming an *accretion disk*. Frictionlike forces on the gas cause the orbiting material to heat up, gradually lose energy, and spiral into the black hole.

The key to forming an accretion disk lies in the strong tidal forces generated by the black hole. To illustrate the awesome power of these tides, let's consider the rather dismal scenario in which Earth collides with a black hole. With only a million or so stellar black holes in our galaxy, the chance that a black hole will collide with Earth is far too low to cause much concern. The odds of a direct collision are roughly one part in 10^{26} per year. Nevertheless, we can describe the sequence of events that would transpire if by some monstrously unlucky coincidence a two-solar-mass black hole is on a collision course with Earth.

The intruding black hole formed during the death of a massive star in the disk of our galaxy. It rushes through space toward Earth with a speed of several kilometers per second. The first indirect hints of the encroaching black hole come millennia before the actual collision. As the tiny behemoth passes through the Oort cloud, the diffuse sphere of comets surrounding the Sun, it dislodges comets from their orbits. Some are thrown into deep space, large numbers are enlisted into orbit around the black hole, while others are dispatched toward the inner solar system to lend occasional brilliant displays to the nighttime skies. Centuries before the black hole's arrival, astronomers notice sizable changes in the orbits of the outer planets as they respond to the black hole's gravitational influence. By examining how the planets stray from their predicted orbits, astronomers can deduce the location, the mass, and the speed of the incoming hole. It would be immediately clear that the interloping object is no ordinary star. A star of two solar masses charging through the outer reaches of the Oort cloud would appear as bright as a streetlight a few blocks away. Arguments might arise as to whether we faced the arrival of a black hole or a neutron star, but these squabbles pale next to the realization that an object with twice the Sun's mass is headed for the core of the solar system.

Telescopes trained in the direction of the black hole register bizarre fluctuations in the brightness of background stars and galaxies. Viewed from afar, a black hole acts like a lens because its gravity warps space-time so as to magnify and distort the images of objects lying along the line of sight. Stargazers might take rueful pleasure in the stunning precision afforded by the impinging and uninvited telescope.

Meanwhile, calculations of the black hole's trajectory prove truly alarming. A dreadfully close encounter is assured, with a direct collision squarely within the realm of possibility. As the black hole crosses the orbit of Pluto, the planets engage in wild deviations from their normal orderly courses. In the perverse but mathematically correct simulation shown in the figure opposite, Jupiter and Uranus are captured by the black hole, while Saturn and Neptune are flung out to uncharted realms of interstellar space. These banished planets are destined for lonely travels spanning trillions of years between significant encounters with other solar systems.

From Earth's unfortunate vantage point, the black hole approaches from the opposite side of the Sun. The combined gravity of the Sun and the black hole pull the Earth through a distorted echo of an ordinary year. In a matter of weeks, the Sun draws closer to Earth than at any other time in 4.6 billion years of terrestrial history. Earth has circled evenly around the Sun for nearly half the age of the cosmos. Now, over the course of a few final days, all that orderly clockwork is ruined. A stunned humanity wrestles with its collective fate. As the Sun scorches the continents, Antarctic ice melts with alarming rapidity and swamps the coastal cities. Hurricanes of unprecedented force rage over the seas.

The end falls quickly. Observed through a telescope, the black hole appears like a psychedelic comet, surrounded by faintly glowing gas and warped images of background stars and galaxies. In the final hour, real havoc begins. As the tidal forces of the black hole take hold, the side of Earth facing

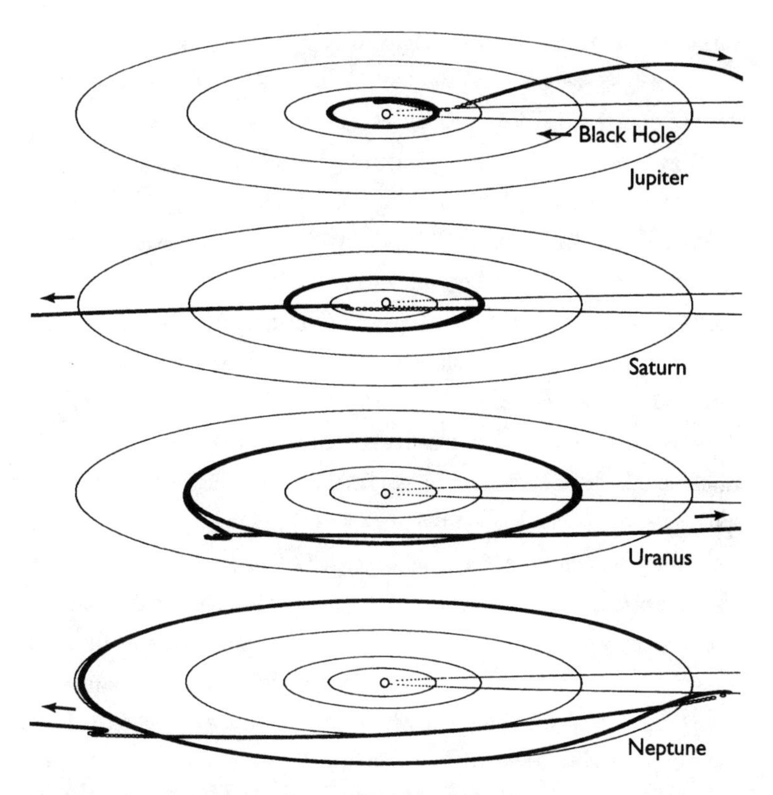

The destruction of the outer solar system by the close passage of a two-solar-mass black hole. In this computer simulation, covering two hundred years, the black hole plunges in through the orbital plane of the solar system on a direct collision course with Earth. Each of the four diagrams shows the disruptive effect of the impinging black hole on the orbital motion of a particular outer planet. For clarity, the action is superimposed on the original orbital paths. In this particular simulation, Saturn and Neptune are thrown from the solar system, whereas Jupiter and Uranus are captured into elliptical orbits around the interloping black hole. Successive orbital positions on the paths are plotted at two-week intervals.

the black hole is pulled with greater force, and the planet becomes grossly deformed. Stresses mount in the crust and earthquakes rock the surface. Six, seven, eight, nine, . . . the Richter scale is outstripped by stupendous temblors. Tsunamis wash over the continents. The planetary crust rips apart along old fault lines, and patches of solid rock float on the scorching lava of the deforming planet. Earth is pulled like taffy into a disk of vaporized rock which forms a whirlpool in its stampede to enter the impinging black hole. The energy released during the final demise is visible far beyond the galaxy.

This collision scenario is melodramatic because of our assumption of a wildly improbable direct hit, and because of our particular concern for the accreted body. But Earth's fictional demise by black hole accretion is an all too real fate for many stars and probably more than a few planets. Indeed, the nearby black hole Cygnus X-1 is now accreting material from its companion—a blue supergiant star. The supergiant and the black hole are locked in a tight orbit, and the star is severely distended by the tidal force of the black hole. A streamer of material is continually siphoned off the star, and forms a whirlpool-like disk of hot glowing gas. Friction causes the gas to heat up and spiral inward, thereby feeding the black hole. The resulting disk scintillates with X rays. Although the blue supergiant itself is slated to explode in a supernova, the black hole will consume a sizable fraction of the star in the interim.

Nascent supermassive black holes at the centers of galaxies are the most prodigious stellar consumers. Smaller forerunners to these supermassive holes form in the dense cores of newborn galaxies, where a great deal of raw material is available to feed the holes. Some of the gas near the black hole has too much angular momentum to fall directly onto the hole. This gas forms a disk around the massive black hole, much like the disk around Cygnus X-1, except on a larger scale. Entire stars that venture too near the hole are torn apart by tidal forces and added to the disk. We can observe the resulting tremendous release of energy as a quasar.

GRAVITATIONAL RADIATION IN THE BLACK HOLE ERA

As the Black Hole Era progresses, energy is slowly drained out of orbiting astronomical systems due to the dissipative effects of gravitational radiation. Since this source of energy loss is hardly a defining feature of everyday life, let's begin with an analogy. Imagine stretching a thin elastic sheet across a metal hoop and then placing a heavy ball on the center of the sheet. The ball sinks down and forms a depression. Next, think about how a marble would roll on the sheet. If started with the proper push, the marble rolls in a circle around the heavy ball. Over the course of a few orbits, it spirals inward, and eventually winds up at rest near the bottom of the depression. The kinetic energy of the marble is lost to the frictional forces between the marble and the sheet; the energy is dissipated as heat and sound waves.

A similar phenomenon occurs when black holes or stars orbit each other. Mass causes space to become locally curved. In our embedding diagrams, the curved two-dimensional surface—analogous to the surface of the elastic sheet—illustrates how distances are organized between points in curved space. Near a star, the curvature of space allows the orbit of a planet and is analogous to the curvature in the sheet. The planet-star system also contains a feature that is analogous to the friction between the sheet and the marble—this friction caused the marble to spiral inwards. Indeed, a frictionlike effect due to gravity does exist. When massive bodies move around or oscillate, they create extremely weak disturbances or ripples in the underlying structure of space-time. As a massive body moves through space, the curvature of space-time must also move, and the movement sends disturbances traveling outwards through space-time. These disturbances are gravitational waves. They carry energy away from massive gravitating bodies in much the same way that vibrations in the elastic sheet and in the air carry energy away from the rolling marble.

Another analogous process allows radio transmitters to produce radio broadcasts. In a radio transmitter, charged particles (electrons) move around and thereby create disturbances in the electromagnetic fields. These disturbances are the radio waves that carry music, sports, and weather broadcasts around our planet. Whereas electromagnetic radiation arises from the movement of electric charges and hence from electromagnetic forces, gravitational radiation arises from the movement of masses and the force of gravity.

Because gravity is so weak, the amount of energy carried in gravitational waves is generally very small. If one waits long enough, however, the effects of this minuscule energy drain become substantial, and objects in orbit must slowly spiral inward. The decay of orbits via gravitational radiation generally takes place at an excruciatingly slow pace. In the absence of other catastrophes, for example, Earth would take more than 19 cosmological decades to spiral into the Sun.

In spite of its lax attitude, gravitational radiation does drive the orbital decay of one observed astronomical system—the *binary pulsar.* This binary system contains a pair of neutron stars that orbit each other every eight hours. They are separated by a million kilometers, a distance similar to the diameter of our Sun. The neutron stars themselves are only 20 kilometers wide, so they are quite far apart compared to their size. This binary pulsar system provides us with two extraordinarily good clocks. Radio pulses resulting from the spin of one of the neutron stars define one kind of clock. When radio pulses are observed from a neutron star, it is called a pulsar. The regular variations in the pulses that arise from the orbital motion of the two neutron stars provide a second type of clock. Both clocks allow for very precise measurements of time, but when their readings are compared, the orbital clock has a clear tendency to gain time. Gravitational radiation causes the orbit to decay and the distance between the stars is gradually shrinking. This constriction of the orbit, in turn, forces the stars to circle each other more rapidly, which drives the orbital clock ever faster (compared to the pulsar clock).

The orbit of the binary pulsar spins itself up exactly in accordance with the theory of general relativity and the accompanying idea that gravitational radiation leads to orbital decay. The binary pulsar thus allows us to measure the predictions of general relativity extremely accurately. Joe Taylor and Russell Hulse discovered this system in 1974 and performed careful timing measurements over the next two decades. Their results are in stunning agreement with the predictions of general relativity and they were awarded the 1993 Nobel Prize in physics.

The binary pulsar will not last very long compared to the long time scales to which we have become inured. In a mere 250 billion years, the two neutron stars will finish their inward spiral and merge. Their gravitational coalescence creates a new black hole, along with an enormous burst of gravitational radiation. The binary pulsar will thus be gone long before the end of the Stelliferous Era. The ensuing black hole, however, is far more durable and should survive well into the 65th cosmological decade.

As the universe drifts into the Black Hole Era, the only remaining binary "star" systems are those whose components are black holes. Just as the binary pulsar tightens its orbit, all binaries experience orbital decay and eventual merging events, replete with the fireworks of gravitational radiation. In order to survive into the Black Hole Era, a binary pair must begin with a separation greater than two light-years. To survive longer, binaries need larger starting separations to outlast the inexorable drain of gravitational radiation. By the 50th cosmological decade, surviving binaries must be initially separated by 650 light-years; to outlast the 60th cosmological decade, binaries need a separation larger than the size of the Milky Way galaxy. To last until the 80th cosmological decade, a black hole binary needs an initial separation larger than the entire observable universe of today.

HAWKING RADIATION AND THE DECAY OF BLACK HOLES

Black holes are not completely black. Over vast expanses of time, they radiate heat into space at an extraordinarily slow rate. Heat is a form of energy, and energy is equivalent to mass. So an object that generates heat must also be slowly losing mass. As a black hole leaks its mass energy away, its rate of heat loss gradually mounts, and hence the black hole cannot last forever. It is destined to evaporate into nothingness.

It seems contradictory that an "object whose surface gravity prevents even light from escaping" can nevertheless radiate energy into space. Indeed, the phenomenon of black hole radiation is a quantum mechanical effect that cannot be understood by using the theory of general relativity alone. Black hole evaporation is one of the few calculable results that fall in the realm of *quantum gravity.*

The quantum mechanical process that causes black hole evaporation was discovered by Stephen Hawking in 1974, and has been named *Hawking radiation* in his honor. The Hawking radiation streaming from the surface of a stellar or supermassive black hole consists mainly of photons and neutrinos, along with a smaller fraction of gravitons. The spectrum of radiation emerging from a black hole has exactly the same form as the blackbody spectrum of photons emitted by an object with a single fixed temperature (as discussed in Chapter 1). The temperature of a particular black hole is determined by its mass; the larger the black hole, the lower the temperature. A black hole with the mass of the Sun has a temperature of only 0.0000001 degrees above absolute zero, while supermassive black holes are vastly colder still.

The radiation of heat from a black hole can be explained with Heisenberg's quantum uncertainty principle. In exploring the properties of white dwarf stars, we saw that the uncertainty principle prevents the momentum and the position of a particle from both being precisely known at the same time. In an analogous fashion, the uncertainty principle also implies that the energy of a system and the time interval over which the system contains a particular amount of energy cannot both be precisely known. This energy-time form of Heisenberg's uncertainty principle implies that the law of energy conservation holds in an average sense, rather than in an absolutely exact sense. Energy conservation can be violated, provided that the violation occurs over a sufficiently short time interval. Larger violations of energy conservation take place over shorter time periods. Pairs of particles, called *virtual particles* because of their temporary status, are continually being created from the quantum fabric of space. These particles live for only a very short span of time, after which they must annihilate back into nothingness. One way to envision this unusual concept of virtual particles is to imagine that the vacuum—supposedly empty space—has the capability to loan out energy. Pairs of virtual particles obtain energy loans and then live on borrowed time for a short while before being compelled to repay their debt to the vacuum and disappear.

To see how this concept relates to Hawking evaporation, imagine a pair of virtual particles created near the event horizon of a black hole. During the pair's brief moments of existence, it is possible for one member of the pair to fall into the black hole and thereby gain energy. If enough energy is gained in this manner, the particles find themselves promoted from their temporary virtual status to the firmament of real existence. To achieve this transformation, a small portion of the gravitational energy of the black hole is used to create the mass energy of the particles. Having been granted a bona fide existence, the second particle of the pair, the one that did not initially fall into the black hole, is free to escape. In many cases, this newly created particle will fall into the black hole and completely repay the black hole for its energy sponsorship. In other instances, however, the second particle escapes from the clutches of the hole and flys off into space. After such an escape, the mass-energy of the black hole is left slightly depleted. The black hole has radiated energy into the background universe.

An alternative way to visualize particle production by the Hawking effect is through the enormously strong tidal forces near the surface of a black hole. When virtual particle pairs are created, these strong tidal forces can pull them apart so that they can no longer annihilate with each other. The tidal force does work on the pair of particles and thereby gives them energy. If enough energy is bestowed upon the pair, that is, if the tidal force performs enough work, then a virtual particle can be promoted in status to a real particle that carries away energy that originally belonged to the black hole.

In any case, the net result of the Hawking process is for the black hole to emit radiation and particles, which carry energy away. In this manner, the black hole slowly loses its energy and slowly loses its mass. Over sufficiently long expanses of time, the energy drain caused by Hawking radiation takes its toll, and the black hole gradually evaporates.

Thermodynamics dictates that heat must flow from hot to cold, but not the other way. Since black holes radiate heat (energy), they must also have a temperature. Every black hole is slightly hotter than absolute zero—the temperature of cold empty space. These black hole temperatures are incredibly small: a black hole with the mass of the Sun has a temperature of only one ten-millionth (10^{-7}) of a degree kelvin. Larger black holes hold onto their mass energy more tightly, radiate much less efficiently, and have even lower temperatures. The surface temperature of a black hole is inversely proportional to its mass. With three million solar masses, the black hole in the center of our galaxy has an effective temperature less than 10^{-13} degrees kelvin. The largest black holes, those with billions of solar masses living in the centers of active galaxies, have even lower temperatures.

The universe is immersed in a bath of radiation left over from the Primordial Era, just after the big bang. This cosmic background radiation gives the universe an effective temperature, which is presently about 3 degrees kelvin.

At this temperature, the universe is considerably hotter than black holes, and heat is flowing from the universe into the black holes at this current time in cosmic history. Although the effect is relatively small, black holes are actually getting more massive by absorbing radiation. As the universe ages, however, the background radiation fields become stretched to longer wavelengths and the effective temperature decreases. At some time in the future, when the universe eventually becomes cool enough, the black holes will lose energy and mass to the background universe. The time at which this transition occurs depends on how fast the universe is expanding—whether the universe is open or closed—and on the mass of the black holes.

If the universe is flat, the temperature of the sky becomes colder than the temperature of a one-solar-mass black hole during the 21st cosmological decade. For black holes of one million solar masses, the transition occurs during the 30th cosmological decade. Similarly, black holes with one billion solar masses begin to evaporate during the 35th cosmological decade. By the beginning of the Black Hole Era, all but the very largest black holes are actively emitting energy and losing mass.

In addition to the cosmic background radiation, the annihilation of dark matter and the decay of protons in white dwarfs also produce seas of radiation that pervade the future universe. These additional radiation backgrounds also delay the evaporation of black holes. Eventually, however, the expansion of the universe stretches these radiation fields to sufficiently low temperatures so that black holes radiate energy faster than they soak it up.

The total lifetime of a black hole depends on its starting mass. Larger black holes have more mass to radiate away, lower temperatures, and last much longer. A black hole with the mass of the Sun evaporates in about 65 cosmological decades, one hundred thousand quadrillion quadrillion quadrillion quadrillion years, a time so long that it almost seems ridiculous to write it out this way. The smallest black holes that are expected to be present in substantial numbers have masses of three to five times that of the Sun and will evaporate away during the 67th cosmological decade.

Black holes with the mass of a million Suns, such as the black hole that resides at the center of our galaxy, outlive the stellar black holes by a substantial stretch of time. In 83 cosmological decades, however, these black holes will also expire. Even a black hole with a mass comparable to a galaxy, one hundred billion solar masses, evaporates through this process by the 98th cosmological decade. As a result, after the 100th cosmological decade, most of the black holes will be gone, and the universe will consist mainly of radiation, neutrinos, electrons, positrons, and other decay products.

The evaporation of black holes through Hawking radiation provides yet another example of the continuing astrophysical war between gravity and thermodynamics. Black holes are a natural consequence of strong gravity, as described by the general theory of relativity. Black holes form when gravity wins its battle against pressure and overwhelms all other forces. On the other hand, the evaporation of black holes is a classic example of entropy production. The radiation produced by this process has a great deal of entropy. The fact that even black holes evaporate means that thermodynamics must eventually prevail, even for these extreme astrophysical objects.

INSIDE BLACK HOLES

What is actually inside a black hole? This question is exceedingly difficult to answer, and lies at the forefront of present-day research on black holes. Nonetheless, this question provides a forum for discussing many fascinating possibilities.

One of the difficulties with talking about the interior of a black hole is the presence of the event horizon, which acts like a one-way membrane. The horizon allows information to pass through into the black hole, but allows no information to come back out. The presence of the horizon, in conjunction with the laws of general relativity, imply that black holes are in one sense rather simple objects. As noted above, they have no hair.

But what does this expression really mean? For one thing, it just doesn't matter what kind of material is used to make the black hole. Once the material is inside the black hole, only the total mass, along with charge and angular momentum, can be discerned from the outside. Protons, iron nuclei, and exotic dark matter particles all have the same effect once they are used to build a black hole. Only their mass, charge, and angular momentum contribute to the black hole properties.

To make these ideas more concrete, let's consider an ordinary star and conceptually allow it to contract into a black hole. The star will have a great deal of hair, complicated structures such as magnetic fields, flaring activity, surface deformations, and star spots. Now let the star implode to form a black hole. The resulting black hole has no way to keep its magnetic fields, flares, and other structures. Loosely speaking, if the black hole could hold onto such features, there would be a link between the inside of the event horizon and the outside. Since such communication is strictly forbidden, the black hole must relinquish all of its extraneous properties. In practice, the black hole does so by radiating away all of the excess energy associated with these structures. At the end of the day, the only properties remaining are the black hole's mass, charge, and spin (angular momentum). All other memories of the original star must be erased.

All black holes with the same mass, charge, and spin are thus *exactly the same.* Because of this elegance and simplicity, and despite their exotic nature, black holes are the stellar objects that are best described by theory. Only three numbers determine black hole properties, and the theory of general relativity includes all three. This state of affairs stands in stark contrast with the case of ordinary stars like our Sun. We have a good theoretical understanding of stars and we can predict their lifetimes and general properties. But stars are so intrinsically complicated that we can never write down equations that describe every stellar property—every stellar flare, star spot, and coronal loop. One requires a great deal more than three numbers to completely describe a star.

Now let's consider the black holes of the future. These black holes can and will interact with the outside universe, especially by emitting Hawking radiation. The emission of radiation from any stellar object, including a black hole, can either make the object spin faster or slower. Which option takes place depends on the manner in which the radiation is emitted (what angular momentum it carries away). For black holes, the emitted radiation tends to slow down the rotation, rather than speed it up. As a result, the spin of black holes is the first to go. The black holes of the far distant future will thus be left with only two properties—their mass and charge. For large black holes, those formed by astrophysical processes and sufficiently long-lived to survive into the Black Hole Era, the charge is very much smaller than the mass. Most of the time, the black holes are characterized by a single number—their mass. The charge plays a role only in the final moments of the black hole evaporation process.

The "no hair conjecture" implies that we cannot communicate with the properties inside the event horizon of a black hole, except for the mass, charge, and spin. But *something* must be inside! In order to discuss the real interior of the black hole, we must use what can be termed "pure theory." In other words, we use the laws of physics—here mostly Einstein's theory of general relativity—to calculate what goes on within the black hole, but we have no way of directly testing these ideas.

According to classical general relativity, without the inclusion of quantum mechanical effects, the most dramatic property of a black hole interior is its central *singularity* in the space-time. Roger Penrose, a renowned theoretical physicist and mathematician, proved a theorem which shows that every black hole must contain a singularity. This result is far reaching and holds independently of how the black hole was made and any other historical artifacts. A singularity is a point where, roughly speaking, all hell breaks loose. At the location of the singularity in a black hole, the density of the material becomes infinite. By infinite, we do not mean just a very large value, but really a value that is greater than any number you can imagine. Clearly something interesting must occur at a singularity.

Usually, when physical quantities become infinite in a theory, it means that something has gone woefully wrong or the theory is incomplete. The singularity at the center of a black hole arises because our understanding of physics remains inadequate at sufficiently high energies and high densities, or equivalently, small size scales. For the enormous densities near the putative singularity inside a black hole, quantum mechanical effects must play a role. In spite of this necessity, however, we do not have a complete and self-consistent description of physical laws that simultaneously include both gravity (general relativity) and quantum mechanics. Thus, while quantum gravity undoubtedly determines the true nature of the singularity within a black hole, a definitive description is currently unavailable.

Nevertheless, a host of intriguing possible effects can occur at the singularity. Some physicists have conjectured that the singularity provides a route from our universe to alternate universes, or perhaps to a different place within our universe. We must be careful about what we mean by our universe and other universes. In this context, our universe is the entire region of space-time that is causally connected. In other words, if you had a spacecraft that could travel at light speed, and you had the entire age of the universe to travel, our universe would contain all of the places that you could visit. If you were to travel inside the event horizon of a black hole, you would never be able to return to our universe. The event horizon thus represents an effective boundary to the universe. But, in principle, something can travel to the singularity at the center of a black hole and emerge in a different universe—a universe with a space-time that is not connected to our universe, except perhaps at the singularity of the black hole. In this manner, the singularities within black holes can provide gateways to other universes.

The very existence of black holes within our universe necessarily implies that the space-time of our universe does not have a simple geometry. In addition to the space-time curvature, produced in conjunction with the strong gravity of black holes, the event horizons provide effective boundaries of our universe.

COMPLEXITY IN THE BLACK HOLE ERA

From a strictly reductionist viewpoint, a human being is a large collection of protons, neutrons, and electrons. These three basic components constitute atoms, which cohere to form molecules, which are organized in a fantastically complex manner to form cells, and trillions of cells cooperate to make a person. In a marvelous way, almost 10^{29} simple particles interact to produce a system that seems to be much more than the sum of its parts.

Our world is complex because there has been a great deal of time for the huge numbers of protons, neutrons, and electrons to interact and evolve into interesting structures. Planets have formed, oceans have laid down sediments, and life has evolved because the universe has existed much longer than the nanosecond time spans required for chemical reactions to take place. If we were to travel back in time to the epoch of nucleosynthesis, when the universe was only a few minutes old, it would be difficult to imagine how the nearly uniform sea of hydrogen and helium nuclei could be endowed with the capacity to form anything as complex as a simple computer—much less a planetary society of five billion persons interacting in an ecosystem of astonishing richness and detail.

Clearly, a large number of particles which follow simple well-defined laws can act in concert to form very complex structures. We want to give some flavor to the idea that the multitude of black holes in the Black Hole Era might assume the role that protons, electrons, and neutrons fill today. Is it possible, given enough time, enough space, and enough black holes, that genuinely complex structures can develop? Can a world made from interacting black holes exist in the same sense that our world of protons, neutrons, and electrons exists? We don't know the answer, but it seems within the realm of possibility. In particular, we can explicitly describe how simple analog and digital circuits can be constructed out of interacting black holes. With circuits in hand, one can readily imagine building computers. And if computing machines are possible, life and intelligence might not be far behind.

Black Hole Computers

Fantastic as it sounds, a collection of black holes, a self-gravitating system, might function as a type of computer. As we delve into this chain of reasoning, we sidestep the obvious question of *how* such a computer might be constructed in practice. That is, we will not worry about how the black holes are placed in their required orbits, or how large masses can be situated to produce the required background gravitational forces that maintain the structure of our hypothetical computer. What we demonstrate, however, is that once appropriate collections of black holes are in the proper configuration, a viable black hole computer could indeed function. The components of our putative, and purely theoretical, black hole computer are certainly not the most efficient nor the most practical devices that could be designed. Nevertheless, if these components were assembled, they would work.

At the most basic level, digital computers are built from three fundamental logic elements, which are usually called NOT, AND, and OR gates.

By combining large numbers of these simple gates, which perform the basic operations of logic, a computer of virtually unlimited complexity can be constructed.

Logic gates operate on numbers, or more precisely, representations of numbers. Any number can be written or represented in binary form, as a series of 1s and 0s,

10101010100001101010 . . .

The number one hundred, for example, can be written as

1100100.

The first challenge in designing a black hole computer is to represent binary numbers using black holes. Perhaps the easiest way to achieve this goal is to use a series of black holes traveling through space as a string of digits. Imagine a line with spaces at regular intervals. Each space in the line may or may not be filled with a black hole. If a black hole is present in a particular slot, then the number we are constructing has a one (1) assigned to that position. On the other hand, if the space is empty, the number has zero (0) assigned to that position. To represent the number one hundred, we need a string of seven spaces with black holes in the third, sixth, and seventh positions (counting from the first digit on the right). Representations of larger numbers require longer strings of black holes and spaces.

Now that we have representations of numbers, we can build logic gates that operate on the numbers. As the first example, we construct an OR gate, which takes two numbers as input data and then produces a single output stream. The two incoming numbers are represented in binary form and can be lined up so that the first digits, the one's place, of each number are side by side. If either of the incoming digit streams has a 1 in a given place, then the output stream has a 1 in that place. For example, let the two input streams be

101000101110

and

010101010101.

After these numbers pass through the OR gate, the output stream, the new number, becomes

11110111111.

In order to enforce this operation using strings of black holes to represent numbers, we must build a gravitational potential well (or field of force) which channels the two streams of black holes together side by side. As the two strings of black holes become closer, gravitational attraction takes hold. When the distance between the two streams becomes much less than the distance between successive spaces within each stream, the black holes (if present in a given space) merge and form new black holes. We have thus made an OR gate. If either of the two input streams has a black hole at a given position, the output stream also has a black hole at that position (see the figure below).

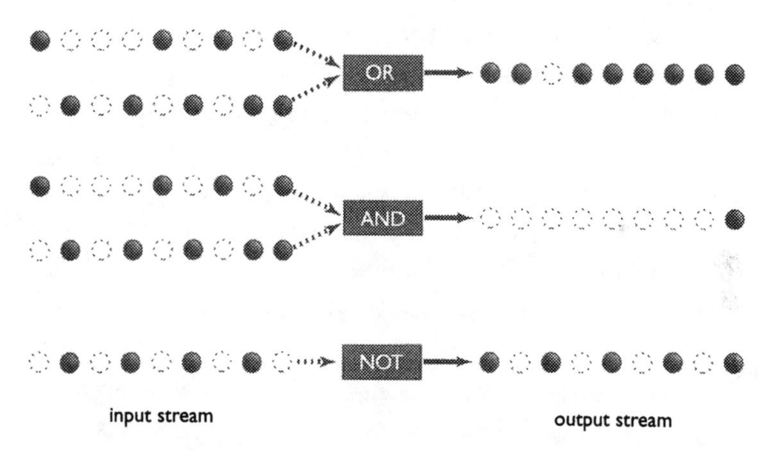

This diagram shows how a black hole computer would work. The top section of the diagram shows a gravitational OR gate. Two input streams of black holes (two "numbers") enter the gate, and a single stream of black holes emerges (a single output "number"). A given position within the output stream contains a black hole if the corresponding position in either of the input streams was occupied. The middle section of the diagram depicts a gravitational AND gate. The output number, again represented by a string of black holes, has a black hole in a given position if and only if both of the incoming digit streams had black holes in that position. The bottom section illustrates a gravitational NOT gate. A single incoming number, a string of black holes, enters the gate, which converts black holes into spaces and spaces into black holes.

Now let's construct the NOT gate. In this logic gate, only one input stream of digits is used. A NOT gate changes all of the digits in the input stream. All of the 1s are converted into 0s and all of the 0s are converted into 1s. For example, under the operation of a NOT gate, the input stream

11010001

is converted into the output stream

00101110.

To construct a NOT gate for strings of black holes, we invoke a rather expensive procedure. The NOT gate itself provides an unbroken stream of black holes, a reference string with a black hole in every space. This reference stream is fired perpendicularly to the input stream, the incoming number. The output stream is the part of the reference stream (not the input stream) remaining after the intersection of the two streams. If a black hole is present in a given position in the input stream, a collision takes place. The total momentum of the collision product removes it from the stream and a space (a 0) is produced in the output stream. Thus, if a black hole comes into the NOT gate, a space goes out. If no black hole is present at a given position in the input stream, the black hole from the reference stream proceeds to the output stream unaffected. Thus, if a space goes into the NOT gate, a black hole goes out. As required, our NOT gate converts black holes into spaces and spaces into black holes (see the figure on the previous page).

The third and final logic gate, called the AND gate, converts two incoming numbers into one output number. The output string of digits contains a 1 if and only if both of the input strings contain 1s in the given place. Otherwise, if one of the incoming streams contains a 0 at the given place, the output stream has a 0 at that place. Using the same two input strings as before,

101000101110

and

010101010101,

the AND gate produces the output stream (number)

000000000100.

To make an AND gate using black holes, we start with the concept of the NOT gate constructed earlier. The first input stream of black holes is passed through a NOT gate so that the output stream of that interaction is "the opposite" of the original input stream. This processed stream is then put into a collision course with the second input stream. The remaining portion of the processed stream, after the collisional interaction with the second input stream, becomes the output stream of the whole AND gate (see the figure on page 141).

Let's see how this gate works. Consider a given position in the stream. If input stream number one has a black hole in that position, then its processed opposite has a space at that position. This space then interacts with the second input stream. If the second input stream also contains a black hole, the output stream will have a black hole. Both streams must thus have a black hole in the given position in order for the output stream to have a black hole.

Although these operations are simple, with enough logic gates a computing machine of enormous complexity can be built. In principle. In practice, a black hole computer constructed from these logic elements will be compromised by three important factors: instability, dissipation, and the evaporation of the components themselves. Instabilities drive the overall disruption of the system due to its self-interaction. Dissipation leads to loss of energy and the distortion of black hole orbits. Finally, the black holes themselves have a long, but finite lifetime. When they evaporate, the computer will clearly cease to compute.

Like our black hole computer, systems made from objects that interact via gravity are often unstable. Consider, for example, the stock science fiction scenario in which our solar system harbors an evil planet on the far side of the Sun. This evil planet is purported to have exactly the same orbit as Earth, but is displaced by exactly one half of a year. The two planets never see each other because the Sun is in the way. This configuration is unstable, however, and this scenario is untenable. Imagine moving the Sun slightly away from the center of such a system (see the figure below): both planets cooperate to pull the Sun farther away from the center. You simply can't keep the Sun balanced between the opposing gravitational forces. In the absence of careful design, our black hole computer is also subject to instabilities of this sort. If one of the black hole "digits" in our binary numbers is shifted slightly from its proper place, the other holes down the line may pull it further from its proper position, perhaps leading to an erroneous calculation, or worse yet the destruction of the entire number. To push back the time required for such instabilities to wreak havoc on our machine, we can make the computer larger and keep the black holes farther apart. It might also be possible to devise more clever and complicated logic elements that are less prone to gravitational instabilities.

In addition to the instabilities which plague the logic components, our black hole computer, like any physical system, is vulnerable to various kinds of

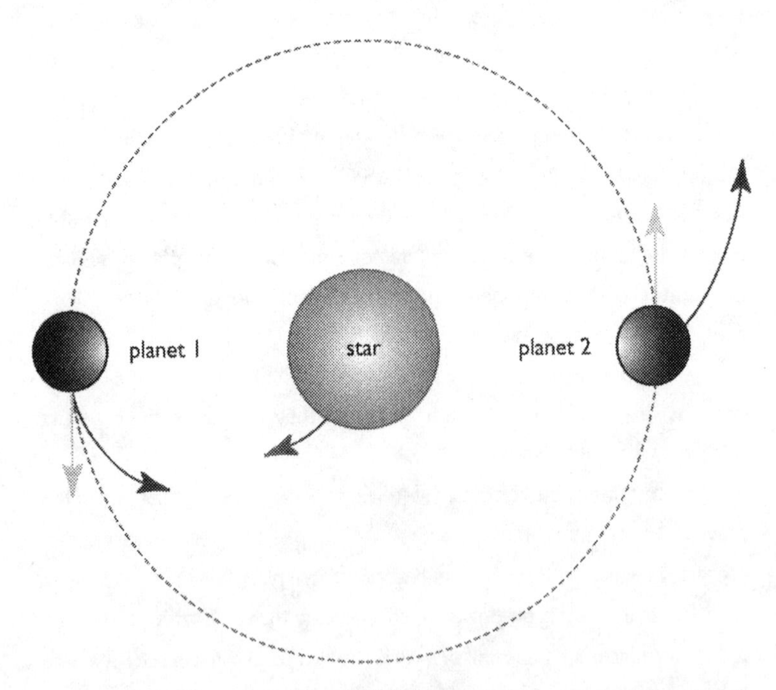

If two planets in the same orbit occurred in nature, the system would be unstable. The combined gravitational pull of both planets would force the central star from the center of the system, and send the planets careening into complicated, unstable orbits.

dissipation. In everyday systems, friction is a common source of dissipation. It leads to machines slowing down, stopping, or wearing out. Within a black hole computer, one obvious source of dissipation is the loss of energy due to gravitational radiation. Moving massive bodies, such as the black holes that comprise our computer, radiate energy as they move through space. As energy is lost, the orbits of the bodies must change in response. The whole system can maintain its integrity only as long as this radiation does not change the orbits too much. Fortunately, as we make the distance between black holes longer to delay instabilities, the time required for gravitational radiation to affect the system also becomes longer. Of course, the time required for the computer to operate also increases—the black hole streams have farther to travel. For a given increase in distance, the effects of gravitational radiation become smaller more rapidly than the operation time becomes longer.

A final obstacle in constructing a black hole computer is that the black holes have a finite lifetime. Given enough time, black holes evaporate away. Although this time is rather long, Hawking radiation limits the amount of computing power that can be obtained in the future.

Black Hole Circuitry

Another way to imagine complex structures in the Black Hole Era is to consider self-gravitating circuits, which can also be called self-gravitating machines. Such circuits are directly related to computers—they can perform analog computations themselves and are necessary to complete most digital computers as well.

A simple circuit, known as an *LRC* circuit, contains three basic components. The first, the inductor, provides an inertia to the system and thus acts somewhat like a mass. The second component, the resistor, dissipates energy and thus provides a frictionlike effect. The third component, called a capacitor, provides a way to store charge and hence store energy. From these simple parts, one can easily construct an oscillator, a device which produces an electric current that varies with time in a well-defined manner. More complex devices can be built from these oscillators and other components.

We can build a simple oscillator from gravitating components available in the Black Hole Era. For the analog of the capacitor, a mechanism to store energy, we use a self-gravitating halo of orbiting bodies much like the halo of our galaxy today. In the future, this halo must primarily consist of black holes. In order to make the energy storage scheme of the halo exactly analogous to a simple capacitor, the halo must have a different shape than that of our galaxy. However, a wide variety of halos allow for oscillations of some type. The remains of galaxy clusters and superclusters constitute large gravitationally bound collections of black holes, just the type of structure required for this halo. Next, we need a large mass—either a large black hole or a collection of black holes bound together into an astronomical system—to actually do the oscillating. The mass of this entity plays the role of the inductance in the circuit. Finally, we need a resistor, a mechanism to provide dissipation. If the masses of the bodies in the halo are small compared to the oscillator mass, then gravitational interactions tend to slow down the larger oscillating body as it moves through the sea of smaller bodies. This effect is known as dynamical friction and provides a resistance. We thus have all of the components necessary to build a simple analog circuit in the future. Furthermore, the components are made from the self-gravitating bodies available during the Black Hole Era.

LIFE IN THE BLACK HOLE ERA

So, if complex machines can exist in the far distant future of the Black Hole Era, what about life? Can living beings of any kind exist in such a foreign environment? In order to contemplate life in this remote setting, we must adopt the optimistic point of view that only the basic architecture for life is fundamentally necessary, and not the actual matter that makes up known life forms on Earth. In particular, carbon-based life is simply not a viable option in this future epoch after all the protons, and hence all the carbon nuclei, have decayed into smaller particles. Life must take other, far less conventional, forms.

Assuming that abstract life forms can exist, we can make some general statements about the nature of such life. According to the scaling hypothesis put forth by Freeman Dyson and outlined in the Introduction, the metabolic rate of an abstract creature is proportional to its operating temperature. For a sentient being, the rate of consciousness, the rate at which the creature experiences life events, scales in a similar way.

The largest temperature that is readily accessible during this future epoch will be the effective surface temperatures of stellar black holes. Due to the Hawking effect, these objects radiate with a temperature of about one ten-millionth of a degree kelvin, approximately three billion times smaller than the operating temperature of a human being (335 degrees kelvin). The largest possible operating temperature for a creature of the Black Hole Era is thus a factor of several billion smaller than the temperature of life forms on Earth. Given this temperature constraint, the fastest conceivable rate of consciousness is slower than that of today by a factor of several billion.

This slowing down of metabolic rates and thought processes is more than compensated by the huge increase in the amount of available time. From the beginning of the Black Hole Era, life forms of the future have a factor of 10^{30} more time than those living in the universe of today. There will be plenty of time for conscious thought, even if it occurs at a much slower rate. For example, this sentence might take a few millennia, or even much longer, to articulate.

Although the slower rates of consciousness are easily accommodated by the larger amount of available time, considerations of energy and entropy place additional constraints on possible life forms of the future. Again following Dyson, we can define the effective complexity of a living creature as its rate of producing entropy, per unit of subjective time. Entropy provides a measure of the amount of information contained in a physical system or process. The subjective time is just the real physical time scaled by the temperature, which takes into account the slowing down of both metabolic rates and thinking rates for creatures operating at low temperatures. This measure of complexity thus represents the rate at which a living creature can process information.

According to this measurement scheme, a human being has an effective complexity value of 10^{23}. To obtain this value, we use the power output of about 200 watts for a person operating at a temperature of about 300 degrees kelvin, and assume that one moment of consciousness corresponds to one second of real time. For purposes of comparison, suppose that we consider a stellar black hole to be a living creature. It would have an effective complexity value of only about 10^{13}. In this sense, human beings are vastly more complex than black holes, by a factor of ten billion. The ramification of this result is profound: Even if some putative life form could harness all of the power of a stellar black hole, its total complexity would be severely limited compared to life today.

THE FINAL MOMENTS

A black hole's last seconds are dramatic. As a black hole shrinks in mass and size, its temperature and evaporation rate gradually increase. When a black hole shrinks to the mass of a large asteroid, its Hawking temperature is similar to room temperature, and it shines faintly in the infrared. After shedding 95 percent more of its mass, the black hole surface is as hot as the Sun. A solar-temperature black hole is an interesting object, especially in this dim penultimate era. With a mass of 10^{22} grams, its gravity would not seem particularly fierce unless you ventured reasonably close. If you hovered 10 kilometers above the event horizon, the force of gravity from the hole would be slightly less than the force of gravity at Earth's surface. From this 10 kilometer distance, the solar temperature hole would resemble a dim star in the night sky. The star would be invisible to the naked eye, but could be seen with a good telescope as a pale white dot floating in the darkness.

Even a relatively hot black hole lasts for a long time. A solar-temperature black hole, for instance, persists for 10^{32} years. During most of this span, the evaporating hole primarily emits massless particles such as neutrinos and photons. A small admixture of gravitons, the massless particles which mediate the gravitational force, also emerge from the black hole. As the mass of the black hole gradually leaks out into space, both the temperature and the evaporation rate increase. As the end draws near, the black hole becomes blindingly bright, and the final vanishing point arrives with a certifiable explosion. During the last second of its life, a black hole converts nearly a million kilograms of matter into radiative energy. As it expires, the black hole produces more than just massless particles. Heavier particles, including electrons, positrons, protons, and antiprotons emerge from the event horizon. In the final sliver of time, a host of more exotic particles are also produced, including, perhaps, the weakly interacting massive particles that populate the galactic halos of today.

The explosive yield of a black hole's last second is a billion times more powerful than the Hiroshima bomb. The resulting blast produces enough energy, mostly in the form of gamma rays, to be visible many light years away. Astronomers have searched the skies for such bursts of gamma radiation, and have found no evidence for black hole explosions. We are thus relatively confident that very few (if any) small black holes exist today. The universe must patiently wait 67 cosmological decades for stellar black holes to squander their mass and ultimately produce the first black hole explosions.

Most of the extremely heavy particles which burst into existence with the death of a black hole are also extremely short-lived, with lifespans far less than a second. These massive particles vanish almost as soon as they are produced. The electrons and positrons emerging from the explosion last far longer. Moreover, the production of protons and antiprotons during the death of a black hole will spark localized revivals of the physical processes that involve ordinary baryonic matter. Since protons and antiprotons are generated in nearly equal numbers, the departure of a black hole is marked by an afterglow of gamma rays resulting from matter-antimatter annihilation. A short while later, after the crackle of gamma rays has subsided and the waste products have dispersed, odd chance collections of protons and electrons might give rise to simple chemical reactions, perhaps forming occasional specks of molecular hydrogen ice. These minuscule throwbacks to the high-energy days of the Degenerate Era are doomed to destruction through the same decay mechanisms that sparked the universal proton shortage in the first place. During the 67th cosmological decade, the lifetime of a proton is an inconsequential instant on the universal logarithmic clock. These intermittent flurries of proton physics are but fleeting and transient events, felled to insignificance by ever widening expanses of time.

If a black hole acquires a net electric charge during its lifetime, it can avoid the indignity of total evaporation. Because of charge conservation, an electrically charged black hole is prohibited from radiating its entire mass away. When a black hole becomes small enough that its mass energy is comparable to the electrostatic energy derived from its charge, Hawking evaporation is prematurely arrested. The resulting *extremal* charged black holes apparently have no means to shed their remaining mass. These bizarre nuggets may well be destined to live forever.

Now imagine yourself emerging at a random point in the universe during the 67th cosmological decade, as the stellar black holes gradually approach their final demise. If the geometry of the universe is flat, the typical separation between individual black holes is unfathomably large, about 10^{43} light-years, or 10^{33} times larger than the size of the universe at the present time. Although the energy generated through black hole evaporation is generally insignificant by terrestrial standards, this radiation provides the dominant driving force during the run-down expanse of the Black Hole Era. The dark, dim, and nearly featureless voids are punctuated by occasional explosions registering in the billion kiloton range. These fleeting violent outbursts are separated by nearly incomprehensible spans of space, time, and silence.

WILL THERE ALWAYS BE BLACK HOLES?

If all the black holes evaporate, the universe surrenders one of its major threads of continuity. The explosive end of the last black hole marks a true watershed event. For a few hours, a tiny corner of the cosmos is steeped in bright light. For the last time, if eyes like ours were present, they could actually *see*. As the final high-energy particles from the explosion rush away at the speed of light, the darkness closing over the universe is truly final.

With all the black holes gone, very few fossils remain to hint at the energetic dawn of time. After the Black Hole Era ends, there will be no flashbacks to the previous high-energy eras. The universe has nothing to play the role of supernovae, which light up the Stelliferous Era and revisit the high-energy conditions that prevailed during the previous Primordial Era. The universe will have nothing analogous to brown dwarf collisions, which sparkle during the Degenerate Era to recreate the glories of the Stelliferous Era. The universe will have no outpourings of protons from disappearing black holes which briefly set the Black Hole Era conditions back to the previous epochs.

But will all the black holes really evaporate? The largest black holes in existence today contain several billion solar masses. Left alone, these black holes would disappear by the 100th cosmological decade, and so, on our global time line, the 100th cosmological decade marks the end of the Black Hole Era. It remains possible, however, that black holes can persist far longer than 100 cosmological decades. If black holes continue to merge and gain mass faster than they evaporate by emitting Hawking radiation, the Black Hole Era can stretch to eternity.

In order to know if the Black Hole Era will come to an end, the question of whether the universe is open or flat is vital. A flat universe might always contain black holes. Although a flat universe is destined to expand forever, the expansion continues to slow down as time passes. In the absence of rapid expansion, the contents of far-flung portions of the universe come into gravitational contact and interact. In the distant future of a flat universe, superclusters of dead galaxies attract other superclusters to form ever larger conglomerations of black holes. Within these huge gravitationally bound aggregates, the individual forces between the trillions of black holes that make up the overall melange cause the heavy black holes to sink to the center and merge. The smaller holes are sent careening away from the cluster at high speeds. These processes of structure formation (the production of ever larger black hole conglomerates) and relaxation (the tendency for heavy bodies to sink to the center of a conglomerate) may well continue indefinitely. The net result is that black holes can merge and grow larger more rapidly than they are destroyed through Hawking evaporation.

On the other hand, black hole growth is more seriously inhibited in an open universe. In this case, the universe is also destined to expand forever, but now the expansion rate is much faster. In such a rapidly expanding universe, with the constituent black holes flying further apart at great speeds, it becomes increasingly difficult for the black holes to merge and grow. Although the large black holes in the centers of galaxies may increase their mass by a few factors of ten, further growth is arrested by the rapid expansion. For an open universe, the Black Hole Era must eventually draw to a close, and our current scientific understanding suggests that the transition will be near the 100th cosmological decade when black holes with galactic or supergalactic mass have evaporated. After that time, the remaining dregs of the universe slide onward into the next era.

5

THE DARK ERA

$\eta > 101$

THE NEARLY MORIBUND UNIVERSE STRUGGLES WITH COSMOLOGICAL HEAT DEATH AND FACES THE POSSIBILITY OF UNIVERSALLY TRANSFORMING PHASE TRANSITIONS.

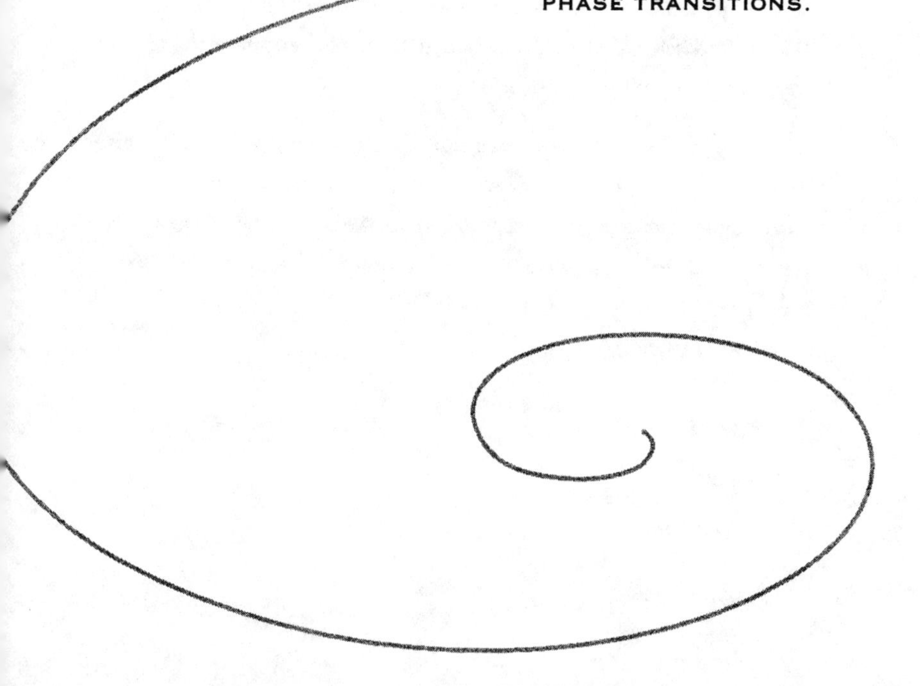

Cosmological decade 185:

Silently, and without warning of any kind, it came. Every cosmic structure it swept over was left disembodied and disfigured in its wake. The destruction was frightening in both its awful swiftness and its devastating completeness.

The shock wave began at a particular but rather undistinguished point of space-time and then traveled outward at blinding speed, rapidly approaching the speed of light. The expanding bubble then enveloped an ever larger portion of the universe. Because of its phenomenal velocity, the shock wave impinged upon regions of space with no advance warning. No light signals, radio waves, or causal communication of any kind could outrun the advancing front and forewarn of the impending doom. Preparation was as impossible as it was futile.

Inside the bubble, the laws of physics and hence the very character of the universe were completely changed. The values of the physical constants, the strengths of the fundamental forces, and the masses of the elementary particles were all different. New physical laws ruled in this Alice-in-Wonderland setting. The old universe, with its old version of the laws of physics, simply ceased to exist.

One could view this death and destruction of the old universe as a cause for concern. Alternately, this natural course of events could be looked upon as a reason for celebration. Inside the bubble, with its new physical laws and the accompanying new possibilities for complexity and structure, the universe has achieved a new beginning.

After black holes have evaporated through the Hawking process, the universe remakes its image once again. As a final enveloping night closes in, perhaps beginning near the 100th cosmological decade, the universe looks vastly different than in any preceding era. In this bleak epoch, the universe is composed only of the smallest types of elementary particles and radiation of extremely low energy and long wavelength. Protons have long since decayed and no ordinary baryonic matter remains.

This low-powered universe bears an uneasy similarity to the very early universe, the first second of cosmic history, when the only cosmic constituents were elementary particles and radiation. In the case of the early universe, however, the background energies were too high for any complex structures, such as stars or even heavy nuclei, to exist. In the far future, the universe contains no complex structures for an entirely different reason: it is so old that all conventional composite entities have decayed away.

Throughout most of cosmic history, the universe was powered by a succession of stellar objects. Energy was first supplied by conventional stars, fueled by nuclear processes. In the next era, degenerate stellar objects ran the show by capturing dark matter particles for annihilation and hence a source of energy. In the end, the degenerate remnants served up their own constituent protons and neutrons for fuel. Finally, the remaining black holes sacrificed their mass energy and evaporated away. After the ultimate demise of this magnificent stellar machinery, the universe must conduct its operations using only the remaining exhaust fumes.

Against this desolate cosmological landscape, the universe faces the possibility of heat death, that is, reaching a static uniform temperature state in which no more interesting events can take place. In spite of the seeming simplicity of this late epoch, however, a host of intriguing possibilities can play themselves out during this cosmic endgame. The cosmological phase transition described in the opening sequence is one possible catastrophe out of the many events waiting to occur as our dying universe staggers into the Dark Era.

SHADOWS OF THE DARK ERA

Taking inventory of the universe at the beginning of the Dark Era, say during the 100th cosmological decade, is an uncertain endeavor. As a general rule, the farther we extrapolate physical law into the future, the less precise our predictions become. Nevertheless, we can make a reasonable estimate of the types and relative abundances of the particles and radiation available at this future epoch. Although we would like to know more, it is quite remarkable that science today can predict anything at all about this future time period so far removed from the present.

ELEMENTARY PARTICLES

Electrons and positrons are key constituents of the Dark Era. Where will these particles come from? To answer this question, we must consider the past history of the universe up to this time. Several different astronomical sources of positrons, electrons, and other particles are available to populate this future epoch.

One important constraint on the future particle inventory is that nature seems to strictly obey a law of conservation of charge. In other words, the universe contains equal numbers of positively charged particles and negatively charged particles. Because of this fundamental symmetry between the positive and the negative, every remaining positron (which has a positive electrical charge) must be matched up with a companion electron somewhere else in the universe.

At the present time, the most familiar type of matter, baryonic matter, is primarily composed of hydrogen. When the proton within a hydrogen atom decays, it often leaves behind a positron. The electron in the hydrogen atom initially remains intact and the net result is thus an electron-positron pair. However, much of the matter in this baryonic class is processed through stars and eventually winds up within the degenerate interiors of white dwarfs and other stellar remnants. As these objects slowly evaporate away through the action of proton decay, the positrons that are left behind find themselves in a dense environment. A thick cloud of electrons surrounds the decay products and thus offers the positrons more than ample opportunity for annihilation. Almost all of the mass energy of ordinary baryonic matter is thus converted into radiation, mostly photons and neutrinos.

Only the leftover protons, those that don't end their lives within stars, are available to make positrons that endure into the future universe. Since the production of stars is not 100 percent efficient, some fraction of the hydrogen and other elements is left behind in a diffuse wisp of gaseous waste. The worthless dregs of one era, however, can be a most precious commodity in a future era. When the protons in this diffuse medium decay, the positrons they sire have a much better chance of avoiding annihilation and surviving into the Dark Era. Even though the majority of the baryonic matter becomes locked up in degenerate stellar remnants, most of the positrons of the future arise from the gaseous refuse left behind by the star-forming process.

The nonbaryonic dark matter of the present-day universe also contributes to the particle inventory of the future. This weakly interacting material currently resides in galactic halos, galaxy clusters, and other large astrophysical structures. A sizable portion of this dark matter will be captured by degenerate stellar remnants as described in Chapter 3. The captured particles annihilate, and their annihilation products are thermalized within the dense stellar interiors. The net result of this process is to convert a substantial fraction of the dark matter mass into radiation, again mostly photons and neutrinos.

The capture of dark matter particles, however, is not 100 percent efficient. A fortunate fraction escape capture and survive into the future. In the long term, these surviving dark matter particles face an uncertain fate. Because we do not know the precise nature of the dark matter, we do not know the lifetime of these AWOL particles. The allowed particle lifetime could be either longer or shorter than the time left before the beginning of the Dark Era, and hence the dark matter particles themselves may or may not survive. Even if the dark matter particles decay, however, their decay products can provide an interesting contribution to the future universe.

Black holes also contribute to the inventory of the Dark Era by spewing particles into space through the Hawking evaporation process described in Chapter 4. This mechanism of destruction converts the bulk of the black hole mass into radiation: mostly neutrinos and photons, with a small component of gravitons. At the very end of the black hole's life, its temperature becomes hot enough that more massive particles are produced as its evaporation accelerates. In particular, the black hole produces electron-positron pairs in substantial numbers. However, the electron and positron production from diffuse hydrogen, gas not processed into stars at the end of the Stelliferous Era, greatly dominates that of black hole evaporation.

In the final moments of a black hole's existence, just before its explosive finale, the surface temperature is hot enough that particles of just about any kind are produced, albeit in relatively small numbers. Black holes thus provide a melange of elementary particles that can survive into the Dark Era. Among the assortment of newly minted massive particles are protons, the building blocks of ordinary matter today. These protons are destined to decay, however, through the same process that marked the end of the Degenerate Era countless years before. As a result, these protons have relatively little impact on the Dark Era.

This glimpse into the future offers an extreme change in perspective. Measured by the clock of a human lifetime, or even the current ten-billion-year age of the universe, the proton lifetime is so long that we usually consider them to live forever. When protons are produced by black hole evaporation, however, the proton lifetime is so short compared to the age of the future universe that they might as well decay instantaneously.

The density of this future era is incredibly low, so low that it is hard to imagine even vaguely, much less fully comprehend. For the sake of clarity, let us focus on the density of positrons. The density of electrons should be exactly the same because physical law enforces charge conservation. Other particles are expected to be even rarer and will have an even lower density.

At the present time, the density of protons in the universe is about one particle per cubic meter. This figure represents a grand average, taking into account all of the protons on extremely large size scales, larger than galaxies. Now suppose that star formation is 99 percent efficient and only about 1 percent of these protons are left over as diffuse gaseous waste. If the universe did not expand, it would be left with about one positron for every 100 cubic meters, a low density to be sure, but one that we can at least envision.

But the universe does expand and it expands quite a lot before the beginning of the Dark Era. For a flat universe, which expands forever but slows down as it does so, the universe grows by a factor of 10^{60} from now until the start of the Dark Era. With this large expansion factor, the future density of the positrons is about one particle for every 10^{182} cubic meters. To get some feeling for the incredible size of this volume, consider that the entire observable universe of today has a volume of "only" about 10^{78} cubic meters. In other words, the positron density of the Dark Era would be about one particle within a volume that is 10^{104} times larger than the universe of today.

For the other possible case, an open universe that expands even faster, the density is much lower still. An open universe grows by a factor of 10^{90} by the beginning of the Dark Era. With this huge expansion factor, 10^{30} times larger than that considered above, the density of an open universe is 10^{90} times smaller than that of the flat universe. A single positron would live within a volume that is 10^{194} times larger than today's universe. No matter how long you think about it, such immensity is difficult to visualize.

The Radiation Backgrounds

Radiation provides another important component of the future universe, and the radiation fields are produced by many different sources. As the cosmos grows older, a series of radiation fields will, in turn, dominate the radiation background of the universe for a range of cosmological decades. Each separate class of radiation is destined to decline in prominence as the universe expands and the constituent photons become successively redshifted into insignificance (see the figure on page 162).

The wavelength of radiation grows longer as the universe expands. This essential feature drives the future evolution and impact of the cosmic radiation backgrounds. One can think of radiation as a collection of "radiation particles" that we call photons. As the universe expands, its volume grows larger, and the number density of photons declines. But the wavelength of the photons grows larger as well and hence the energy of each photon decreases. Because of this additional property of wavelength stretching, also called *redshifting*, photons lose energy density faster than ordinary massive particles in the face of an expanding universe.

At the present time, the cosmic background radiation left over from the big bang is the most energetically interesting and cosmologically important radiation field. This radiation now has an effective temperature of 3 degrees kelvin and characteristic wavelengths of 1 to 2 millimeters. In the future, the wavelength of this radiation becomes severely stretched out as the universe expands. A flat universe, for example, grows by the huge factor of 10^{60} between now and the start of the Dark Era. This expansion stretches the cosmic background radiation to the colossal wavelength of 10^{41} light-years, much larger than the size of the observable universe today.

Other sources of background radiation become increasingly important as the universe ages. Stars are now continually pumping out energy in the form of starlight, whereas the cosmic background radiation is fading from prominence due to the redshift effect. The background sea of stellar radiation will eventually overwhelm the radiation left over from the big bang during the 12th cosmological decade. As the relatively near future unfolds, this radiation will be predominantly produced by red dwarfs, which are the smallest, most common, and longest-lived stars. These relatively cool stars produce radiation with a characteristic wavelength of about one micron, one-millionth of a meter. As the universe expands, this radiation also gets stretched out so that its wavelength grows to nearly 10^{37} light-years by the start of the Dark Era.

The capture and annihilation of dark matter particles in white dwarfs provides another important source of radiation for the future universe. The net effect of this process is to convert a substantial portion of the mass energy of galactic halos into radiation, which becomes the dominant background during the 17th cosmological decade. As this radiation pours out of white dwarf surfaces during the Degenerate Era, its wavelength is about 50 microns, or one-twentieth of a millimeter. These photons also become stretched out as the universe grows ever larger.

The end of the Degenerate Era is marked by the decay of protons and the conversion of ordinary baryonic matter into radiation. With our representative proton lifetime, this source of radiative energy begins its domination of the universal background during the 31st cosmological decade. The characteristic wavelength of this radiation starts at about one inch and then becomes stretched out as the universe continues its relentless expansion.

Finally, the black holes evaporate, ultimately converting their rest mass into photons and neutrinos, and taking over the radiation background some time around the 60th cosmological decade. Stellar mass black holes produce radiation with a characteristic wavelength of several kilometers, comparable to their radial size. Black holes of higher mass have correspondingly lower temperatures and radiate at longer wavelengths. Billion-solar-mass monsters, the black holes that now live at the centers of active galaxies, have characteristic wavelengths of billions of kilometers, about the size of our solar system. All of this radiation, as usual, gets stretched out by the ever continuing expansion of the background space-time of the universe.

HEAT DEATH

Our universe greatly slows down its operative processes as it endures into the Dark Era. But does it ever stop altogether, or slow down so much that it is no longer interesting? Can we reach a time in the future when nothing at all happens? Because of its intimate connection with thermodynamics, the idea of the universe slowing to a stop is known as *heat death.* The possibility of a universal heat death has troubled many philosophers and scientists since the mid-19th century when the second law of thermodynamics was first under-

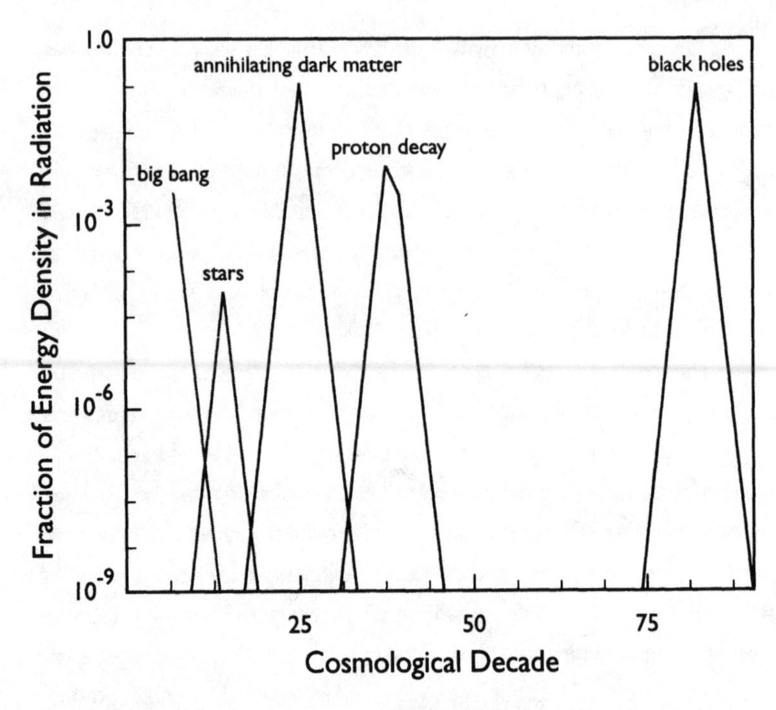

The contributions of various processes to the radiation background of the universe are shown here as a function of time, for cosmological decades 5 to 90. The vertical axis presents the relative energy in radiation from several sources: radiation left over from the primordial universe, starlight, dark matter annihilation, proton decay, and black hole evaporation.

stood. Heat death arguments can take many different forms. We use the term *classical heat death* for a universe that reaches complete thermodynamic equilibrium. In this state, the entire universe has a constant temperature everywhere in space. Without a temperature difference, no heat engine can operate and no work can be done. Without the ability to do physical work, the universe "runs down" and becomes a rather lifeless and inactive place.

How does this heat death come about? The second law of thermodynamics states that the total entropy of a physical system can never decrease (in this case, the system is the whole universe). But entropy is allowed to remain constant, not change with time. The problem is that physical processes that don't produce entropy are usually not very interesting. So, as a general rule, we would like entropy-generating processes to spice up the universe. All physical systems have a tendency to reach a state of thermodynamic equilibrium, which corresponds to a state of maximum entropy. In thermodynamic equilibrium, all parts of a physical system have the same temperature and the entropy remains strictly constant. Interesting processes in the universe will thus shut down if thermodynamic equilibrium is attained.

The universe of today is quite far from thermodynamic equilibrium. The background temperature of the universe is a cool 3 degrees kelvin, about 270 degrees below the freezing point of water (on the Celsius scale). This cold background provides a sharp contrast to the blazing hot surfaces of stars, which have a wide range of temperatures from about 4000 to 40,000 degrees kelvin. This out-of-equilibrium nature of the universe allows interesting processes to occur. Heat flows from the hot stellar surfaces out into space to warm up planets, drive weather systems in planetary atmospheres, and even empower the evolution of life. The universe acts like a gigantic heat engine. The temperature difference is absolutely crucial. If the universe attained thermodynamic equilibrium and achieved a constant temperature everywhere in space, no more work could be done. Interesting processes, like biological evolution, would no longer take place.

A common misconception that arises in discussions of thermodynamics is the apparent paradox of how any complex structures can be produced in the face of the need for ever increasing entropy. After all, entropy is a measure of the disorder of a system. If complex systems are highly ordered, how can they ever arise without violating the tendency for entropy to increase? The resolution of this ostensible paradox is that the total entropy of the system must increase, but the entropy of one portion of the system is allowed to become smaller and hence that one piece can become highly ordered. If one part of the system becomes well ordered and loses entropy, the system as a whole must pay for it by increasing its entropy somewhere else for compensation.

Within the context of modern cosmology, the temperature of the universe is continually changing and the issue of heat death shifts substantially. A continually expanding universe never reaches true thermodynamic equilibrium because it never reaches a constant temperature. The background temperature of the universe continues to decrease because of the expansion.

Classical heat death is thus manifestly avoided. However, the expanding universe can, in principle, become purely adiabatic, which means that the entropy within a given region of the universe remains constant. In this case, the universe can still become a dull and lifeless place with no ability to do physical work. We denote this latter possibility as *cosmological heat death,* which represents an effective heat death, even though the temperature of the universe is not constant. As we discuss throughout this book, interesting cosmological events continue to produce energy and entropy in our universe, at least until after the 100th cosmological decade. Cosmological heat death is thus postponed until the universe evolves well into the Dark Era.

The energy and entropy generating mechanisms available to the universe depend on the mode of long-term evolution. If the universe were closed, it would be slated to eventually recollapse and end its life in a big crunch, and long-term entropy production would not be an issue. The universe would carry on interesting physical processes right up to the final moment of the big crunch. The nomenclature of this discussion is a little bit ironic: A closed universe can escape the indignity of heat death, even as its complex structures evaporate under the intense radiative heat resulting from the catastrophic collapse.

For the case of a flat universe, which slows down as it continues to expand, cosmic structures of ever increasing size and mass enter the cosmological horizon and become bound together through the action of gravity. Because the expansion of the universe slows down, gravity has a chance to pull together material from larger and larger distances as the universe ages. In a flat universe, cosmic structures of huge sizes can be formed well into the Dark Era. Of course, the Dark Era need not be so dark. Some of these enormous cosmic structures can, in principle, collapse to form black holes and hence the previous Black Hole Era might not really end. It is even possible, but certainly not guaranteed, that black holes can form faster than they evaporate. In this case, the universe could continue its operations using the energy produced by Hawking evaporation of these monster black holes. Cosmological heat death can thus be avoided, at least in principle, as long as the universe remains nearly flat. In this case, the war between gravity and thermodynamics reaches an effective stalemate. Gravity continually produces ever larger gravitationally bound structures—black holes—and attains an apparent victory. But each individual structure is destined to evaporate, leading to a compensating triumph of thermodynamics and entropy production.

On the other hand, if the universe is open, the expansion speed approaches a constant value and gravity manifestly loses its battle with the expansion—it can no longer compete. The formation of cosmic structures becomes frozen out at some fixed length scale and the continued formation of black holes, or any cosmic structures, becomes seriously inhibited. In this case, the questions of long-term entropy production and cosmological heat death are still open. Although these prospects may seem rather bleak, many fascinating new possibilities still remain.

LIFE AND DEATH FOR POSITRONIUM

Perhaps the most lively feature of the Dark Era will be processes involving positronium atoms. With no protons or neutrons left, conventional atoms become impossible constructions. On the other hand, positrons, the positively charged antimatter partners of electrons, will be relatively abundant. Electrons and positrons can be cast into atomic structures that are analogous to conventional hydrogen atoms, which consist of one proton and one electron. Atoms made from positrons and electrons are known as positronium.

The atomic properties of positronium are markedly different from conventional atoms in two respects. Because the positron mass is two thousand times smaller than that of the proton, the electron orbits are altered. Positronium chemistry is thus quite different from that of hydrogen. Most importantly, however, the positron and the electron can annihilate with each other, whereas the proton and electron in an ordinary hydrogen atom cannot. The fate of positronium atoms is thus sealed as soon as the atoms are made. Given enough time, the electron and positron must annihilate with one another and produce a tiny burst of radiation.

Positronium atoms are synthesized in terrestrial laboratories in a fairly routine manner. These atoms are generally created in low energy states with microscopic size scales, roughly comparable to the size of ordinary atoms. These microscopic positronium atoms live for only a tiny fraction of a second before annihilation removes the particles from the universe. This short lifetime, which is highly unsatisfying, arises because of the tiny size the atoms are born with.

Fortunately, in the very late universe, the background density is exceedingly diffuse and positronium atoms form with orbits of incredibly large radii. The typical size of positronium forged during the Dark Era is trillions of light-years, larger than the entire observable universe of today. Positronium formation of this type is scheduled to begin around the 71st cosmological decade. These enormous atoms are born in states of relatively high energy, compared to the microscopic positronium atoms that decay so quickly. The electron and the positron slowly circle each other and gradually give up extraordinarily small amounts of radiation as their orbits become ever smaller. The particles spiral together in an exotic dance which ultimately results in the complete destruction of the participants and a total bankruptcy of their stored energy. With their enormous starting sizes, the positronium atoms take a rather long time to decay, about 145 cosmological decades. The future universe thus contains a window of time during which positronium can be made and can live before its inevitable self-destruction takes place. This window is roughly centered on the 100th cosmological decade, just when the black holes with galactic masses are making their explosive exits from the universe.

An important question is whether or not these positronium atoms, or perhaps even more unusual atomic structures of the future, can combine to form complex entities of any kind. Can processes even remotely resembling the chemical reactions that we see on Earth today take place in this dark future? Is 145 cosmological decades enough time for some kind of "biological" evolution to take place? What would life forms of this era look like? These questions remain unanswered, but hold the key to possible life processes during the Dark Era.

The formation and eventual destruction of positronium represent yet another battle in the ongoing war between gravity and thermodynamics, a conflict that continues well into the Dark Era. At this late epoch, the formation of positronium is fundamentally caused by the electrical attraction of the particles, although larger groups of particles can be brought together by gravitational forces. Although extremely long-lived by the standards of today's universe, these positronium atoms are transient structures and are slated to decay into radiation. The inevitable demise of positronium thus represents another victory for thermodynamics and entropy production. Disorder once again triumphs in the end.

THE NEVER-ENDING ANNIHILATION

The simple process of particle annihilation illustrates how the universe slows down, but always keeps going. Annihilation events convert mass energy into radiation and thereby provide an energy source for the universe. In much the same way, the Sun provides an energy source for Earth today, and stars provide an energy source for the universe, although the scale is dramatically different.

The universe of the future, for example, contains both electrons and their antimatter counterpart, positrons. When the particles wander sufficiently close together, they annihilate and release all of their mass energy into a burst of radiation. As this explosive event takes place, entropy is produced. In the case of positrons and electrons, as outlined above, the particles often form positronium atoms before their eventual annihilation. If we view this process on time scales much longer than the positronium lifetime of 145 cosmological decades, however, we need not worry about this transient intermediate step. Other particles, or pairs of particles, can also live into the Dark Era and can annihilate in similar fashion. Weakly interacting dark matter particles, for example, may survive into this time period and may be available for future annihilation events.

Does the universe ever run out of particles to annihilate? The answer is quite illuminating and nicely illustrates the almost perpetual nature of this cosmic endgame. The total fraction of the energy density of the universe that will annihilate during the Dark Era is a small and given quantity. Particle annihilation provides only a finite amount of energy (within a given region of the universe) over the entire time span of the Dark Era. The annihilation rate slows down markedly as the universe expands and grows more diffuse, but particle annihilation continues as long as the universe exists. At no time in the future does the universe reach a state in which no more particles will annihilate. No matter how old the universe becomes, there will always be some explosive annihilation events yet to take place, and to light up the sky in their own small way.

We thus have a slightly confusing state of affairs: particle annihilation continues forever but produces only a fixed amount of energy within a given region of the universe. This apparent paradox is easily resolved by taking into account the decreasing rate of annihilation. The universe effectively doles out a fixed amount of energy over an infinite span of time. This conservation practice underscores the energy crisis of the future and *defines* exactly what it means for the universe to "slow down" as it gets older.

The universe thus slows down dramatically, but never quite runs out of steam. Can one think of the universe, with this never-ending quality, as having an eternal life? Perhaps, but that would be a rather optimistic point of view. Paul Davies, a renowned physicist and writer, has described this late phase of evolution of the universe as an "eternal death." This ever dying universe continues to evolve and continues to slow down, but never reaches a final moment of deathlike closure.

The continuing annihilation of particles, and other related processes, provide another example of the Copernican time principle introduced in the introductory chapter. No matter how old the universe becomes, interesting physical processes continue to operate. In fact, the Dark Era need not be such a dull and boring place after all. Evolution could be sparked by dramatic events of a truly universal character, if enough time is available. But time is the one plentiful commodity in the dying universe of the future.

TUNNELING PROCESSES AND FUTURE PHASE TRANSITIONS

The opening passage of this chapter describes a cosmological phase transition, a potential cosmic catastrophe of stupendous proportions. Given the enormity of this hypothetical but nonetheless possible future event, it is worth examining this process in a bit more detail. The universe could in principle contain a substantial contribution of vacuum energy. In other words, seemingly empty space might not be so empty. Recall that this type of energy contribution to the universe can lead to inflationary expansion, which takes place moments after the big bang when the universe is extremely young. The same type of vacuum energy could be present in the universe today, albeit at a much lower energy density. Once we understand that the vacuum state of the universe is allowed to have some energy, it is not hard to imagine that the vacuum could have many different possible states of differing energy. A universe with multiple vacuum energy states can experience an extremely interesting long-term effect: the vacuum might be unstable and the universe could undergo a future transformation into an entirely new state, one with lower vacuum energy.

Unfortunately, the contribution of the vacuum to the energy density of the universe remains unknown. In fact, the "natural value" of the vacuum energy density appears to be larger than the cosmologically allowed value by many orders of magnitude. In other words, the simplest calculations suggest that the energy density of the vacuum should be larger than the observed total energy density of the universe by a factor of about 10^{122} or so. This incredible discrepancy is generally known as the *cosmological constant problem* and has no currently accepted resolution. The vacuum energy of the universe could be as small as zero or as large as the total energy density due to ordinary baryonic matter, exotic dark matter, and everything else. And we don't know how to reduce this range of possibilities. A successful solution to this rather embarrassing problem will eventually lead to a fundamental step forward in our understanding of the universe: past, present, and future.

For the sake of discussion, let's consider the possibility that the universe does indeed have a vacuum energy density and that the universe is currently in a "false" vacuum state. In other words, the universe is now trapped in a configuration with a "large" vacuum energy, but a lower energy vacuum state exists. In this scenario, the universe can make a transition to the lower energy state through a quantum mechanical tunneling process. As tunneling proceeds, the universe experiences a phase transition, roughly analogous to the transition that occurs when liquid water freezes into solid ice.

The act of quantum mechanical tunneling requires two essential features.

First, the physical system must have more than one energy state so that transitions between them are possible. In addition, the system must possess an energy barrier that makes it difficult for the transitions between different states to take place. This latter feature is important because all physical systems have a tendency to seek their lowest energy state, often called the ground state, just as water flows downhill rather than uphill. Without an energy barrier, physical systems quickly transform themselves into their lowest energy states and remain there for the rest of time. The interesting case arises when a physical system is trapped in a state of higher energy—not the lowest energy state—and can in principle make the transition at some time in the future. The physical system can be an atom, a nucleus, or the vacuum configuration of the universe itself.

The fundamental concept of an energy barrier can be illustrated with a classical analogy. Consider a ball rolling back and forth between two hills as depicted in the figure opposite. With no friction, the ball rolls back and forth in its valley, but never crosses over to the other side because it doesn't have enough energy to get over the hill separating the two valleys. The hill thus provides an energy barrier which prevents the ball from making a transition from one valley to the next. The ball is trapped in the left-hand valley, even though the right-hand valley is deeper and corresponds to a lower energy state of the system.

In the classical example of the ball with two valleys, and a fixed total energy, no transitions are possible. Ever. The ball is destined to remain in its valley forever, unless it is acted upon by some outside mechanism. For a quantum mechanical system, however, the situation is fundamentally different. Because of the wavy aspect of reality on small size scales, nature is never completely still. Physical systems always have fluctuations due to the uncertainty principle that we have met before. And these quantum fluctuations allow seemingly forbidden events to take place.

If the ball in the top panel is replaced by an electron, for example, and the hills are replaced by electrical barriers of the same shape, then we have a completely analogous system, albeit on a vastly smaller size scale where quantum mechanical effects must play a role. Because of quantum fluctuations, the electron always has some probability of being in the right valley, even when it

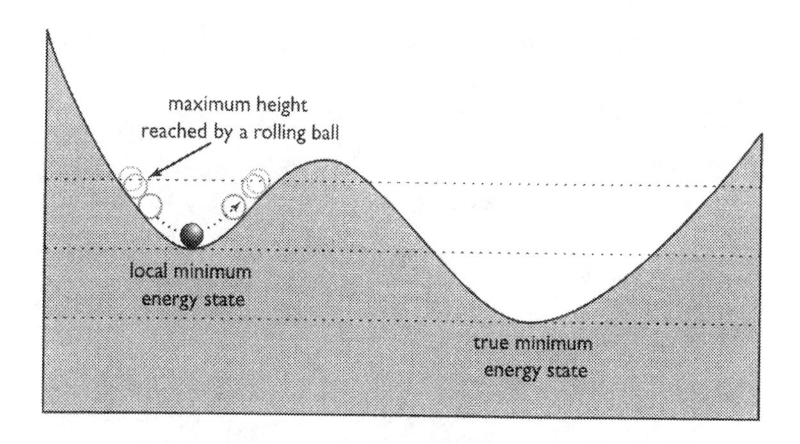

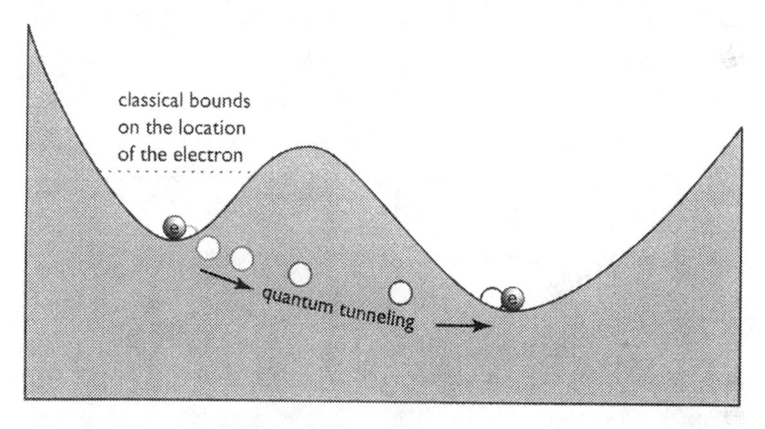

The top panel represents the nature of potential energy in a classical system. In it, a ball is rolling back and forth in the left-hand valley. The right-hand valley is deeper, and hence represents a lower energy state, but the ball does not have enough energy to pass over the hill and into the next valley. The bottom panel represents the nature of potential energy in a quantum system and shows the tunneling of an electron from the left potential well to the right. These potential wells are analogous to the valleys shown in the top panel. The electron doesn't have enough energy to go over the barrier and hence this behavior is classically forbidden. Because of quantum mechanics, however, the electron can make the transition.

is supposed to be in the left valley. The probability of the electron being on the wrong side of the potential barrier is generally quite small, *but it is not zero.* In practical terms, this uncertainty means that, given enough time, the electron will make a transition from the left valley to the lower energy state afforded by the right valley. As it completes its transition, the electron generally gives up energy and then remains in the lower energy valley.

When the electron makes this transition from one valley to the other, it effectively passes through an energy barrier that it did not have enough energy to go over. In this sense, the electron *tunnels* through the barrier and this process is known as quantum mechanical tunneling. This seemingly mysterious behavior is a direct result of the electron exhibiting wavelike properties. Although this quantum behavior of electrons may seem outlandish, electron tunneling provides the fundamental basis for the construction of transistors and other solid state circuit devices. Without the quantum mechanical tunneling of electrons, the entire semiconductor industry would be out of business.

Waves of all kinds, both classical and quantum, are capable of tunneling behavior. For example, spiral density waves, the wavelike disturbances that produce the beautiful spiral patterns seen in galaxies, can undergo tunneling. And these waves are substantially larger than our solar system, safely large enough that quantum effects do not play a role. The strange aspect of quantum mechanics is basically that particles have wavelike properties and exhibit wavelike behavior. Waves of all kinds naturally tunnel through barriers, produce diffraction patterns, and obey an uncertainty principle. Once we accept that particles and other physical systems have wavelike characteristics on sufficiently small size scales, many of the counterintuitive quantum effects can be readily understood.

Like the electron in our example, the vacuum state of the universe could be trapped in a state of high vacuum energy and a state of lower vacuum energy could be available. Once the universe becomes trapped in a state of large vacuum energy, it must stay in that state for an extended period of time because the energy barrier prevents it from immediately transforming itself into the lower energy state. Such a long-lived configuration is called a *metastable state* because it is effectively stable for short periods of time, but ultimately unstable and doomed to decay. Transitions to the lower energy state are not only possible, but must occur if enough time is available.

How, when, and where will such a phase transition take place? We would like to know the probability that the universe will experience a transition from its false vacuum state to the true vacuum state, from a state of high vacuum energy to a configuration of lower energy. For a specific theoretical model of the vacuum state, we can calculate this transition rate in a fairly straightforward manner. The result, however, is extremely sensitive to the input parameters, which are not well known. We need a complete theory of the vacuum state of the universe to make this prediction precise. Although such a theoretical understanding should eventually result from the solution to the cosmological constant problem, it is not currently available. In the meantime, we can only put constraints on the possibilities for this frightening but fascinating future event.

Clearly, the tunneling time scale must be long enough that the universe has not decayed by the present epoch. We can be reasonably sure that no nucleation events have occurred within the observable portion of the universe during the current age of the universe. The time scale for this phase transition can easily be longer than the ten-billion-year age of the universe. In the future, however, the universe could experience a phase transition and tunnel into its lower vacuum state at virtually any time, as soon as tomorrow, or as late as ten thousand cosmological decades from now.

If and when this tunneling event occurs, what actually happens? At a given point in space, a microscopic region spontaneously transforms itself into the new, lower energy vacuum state. When the phase transition begins, microscopic bubbles of the true vacuum state nucleate in the background sea of false vacuum and begin growing outwards. This process is much like the growth of ice crystals in water cooled below the freezing point. The ice crystal starts growing at some well defined, but usually arbitrary point and then grows larger. The icy region expands outwards as an advancing front which converts liquid water into solid ice as it passes. The bubble of the new vacuum state grows outwards in a similar manner and converts the background universe in the old vacuum state into the new phase with a new vacuum state.

The cosmological phase transition takes place rapidly because the bubble walls accelerate extremely quickly. When viewed from large size scales, the bubbles appear to expand at almost the speed of light, almost as soon as they are produced. As the outer wall of a bubble expands, it sweeps up regions of the old universe and leaves behind regions of the new universe in its wake. The bubbles grow until they encounter one another, collide with each other, and thereby complete the phase transition. In this manner, the old universe is transformed into a new universe with a new vacuum state (see the figure below).

Inside the bubble, in the region with a new vacuum state, the universe changes its character almost completely. The physical laws of the universe, including the values of the physical constants, change as the phase transition proceeds to completion. The masses of the fundamental particles, as well as the strengths of the coupling constants that determine the forces of nature, are all different in the new vacuum state. The universe, as we know it, simply ceases to exist.

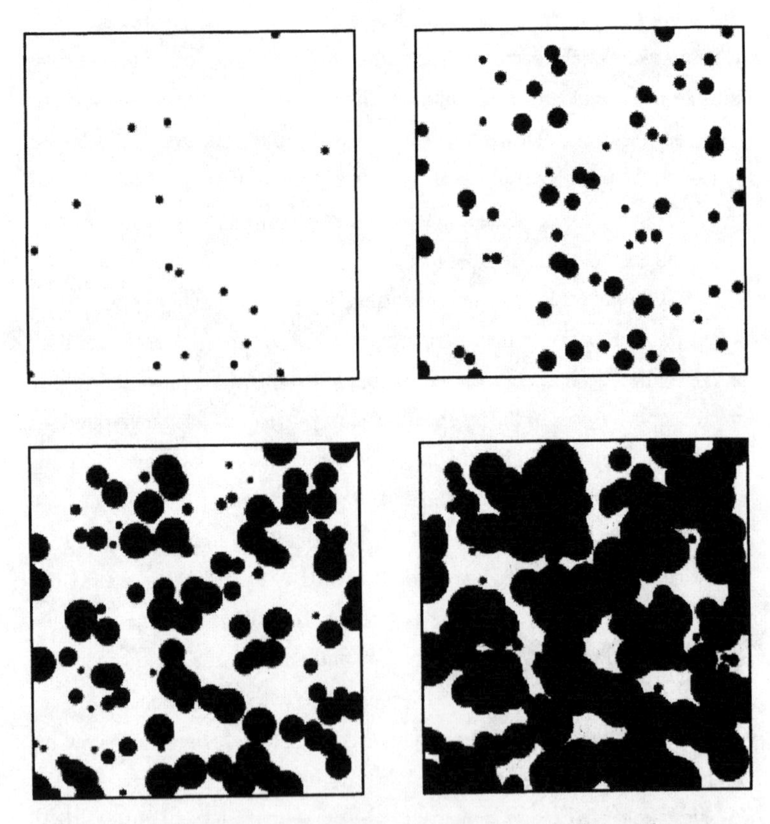

Here we see a cosmological phase transition in progress. The bubbles of the new vacuum state (shaded regions) have nucleated within the background sea of the old vacuum state. The bubbles grow with time until they meet up with each other and the phase transition becomes complete.

Any living creatures caught up in the phase transition would die instantly. They would no longer be capable of their usual biological functions or even any chemical activities. Because of the rapid expansion of the bubbles, at nearly the speed of light, any observers present during this monumental event would have absolutely no warning of its coming. Since no means of communication can travel faster than the speed of light, no advance warning signals could advertise the impending destruction. In short, it would be impossible to see the transition coming.

Once the bubble wave front arrives, the transition time required for the old vacuum state to turn into the new vacuum state is a tiny fraction of a second, from about 10^{-10} to 10^{-30} seconds. This microscopic time scale is considerably shorter than the response time of a neuron in the human brain. The transition is thus far too abrupt for humans to experience. For putative life-forms of the far future, the rates of thought processes should slow down with decreasing temperature according to the Dyson scaling hypothesis. Future life-forms are likely to operate at lower temperatures than humans and should have even slower response times. Any beings witnessing this cosmic phase transition would literally never know what hit them.

The nature of the universe and the laws of physics are markedly different before and after the phase transition, or, equivalently, outside and inside the growing bubbles. The exact nature of the new universe depends critically on the new vacuum state. In particular, the cosmological constant contribution is different before and after the phase transition. The theory of gravity also takes on a new form.

The most optimistic scenario arises when the cosmological constant is currently positive and the universe tunnels into a new vacuum state with exactly zero vacuum energy. In this fortuitous case, the new universe, with its new physical laws, has a chance to develop new starts for complexity and perhaps even new starts for life. For this renaissance universe, the possibilities are rich and varied, virtually unlimited.

On the other hand, if the cosmological constant of today's universe is already exactly zero and the new vacuum state of the universe has a negative value, then a far more serious apocalypse takes place. In this disastrous case, the space-time of the new phase—the regions within the growing bubbles—

is gravitationally unstable. The interior of every bubble suffers a truly catastrophic gravitational collapse. The density and temperature of the interior material increase without bound. The bubble thus collapses into a miniature version of the big crunch, the end-of-the-universe scenario outlined in the next chapter. The bubbles collapse into their fiery death within a few microseconds, or less, far too short for the ascent of biological evolution, or even more modest developments of complexity. This option for the phase transition is thus interesting from a pyrotechnics point of view, but provides little hope for future life.

The quantum mechanical tunneling process that produces the phase transition will happen spontaneously, given enough time. An especially disturbing aspect of this potential disaster is that the phase transition could, in principle anyway, be *induced* to occur. If the vacuum structure of the universe has the proper form, and if the proper technology is available, the phase transition could be started by living beings, human or otherwise. In this consciously directed case, the phase transition begins at a well-defined point in space and travels outward, again approaching the speed of light. A single advancing wave front would sweep up the old universe and leave behind death, destruction, and a new vacuum state in its wake. Once it got started, the advancing front would continue its path of total annihilation, and virtually nothing could be done to stop it. Such an event would represent the ultimate act of terrorism.

Yet another intermediate and alarming possibility exists. A phase transition into a new vacuum state could be triggered accidentally, either by living beings or by natural circumstances. High-speed collisions between cosmic rays with extremely large energies could provide the spark that ignites the phase transition. One can envision other natural disasters as well.

A future phase transition sweeping over the entire universe through the action of quantum tunneling is one of the more speculative topics considered in this book. Nevertheless, its inclusion is appropriate because the act of tunneling from a false vacuum into a true vacuum changes the nature of the universe more dramatically than just about any other physical process. The universe could well be kevorking on the brink of catastrophic instability.

This idea of a cosmological phase transition is not nearly as unusual or unnatural as it might first appear. In the earliest moments of cosmic history, immediately following the big bang, the universe descended through an entire series of phase transitions with vaguely similar characteristics. The first such phase transition led to the incredibly rapid expansion of the inflationary epoch. Another phase transition broke the symmetry of the universe and split the electroweak force of nature into the two parts we now know as electromagnetism and the weak nuclear force. Yet another phase transition altered the infant universe as free quarks condensed into composite particles known as hadrons, including the familiar protons and neutrons we know today. All of this activity, with its profound effects on the character of our universe, took place before the cosmos was even one second old. With almost limitless time available, the future universe has ample opportunity to experience the awesome drama of another phase transition.

THE MAKING OF NEW UNIVERSES

The future universe is capable of supporting even more bizarre behavior. It is even possible for the universe spontaneously to create new "child universes" through a quantum tunneling process roughly analogous to that described above. In this case, a bubble of the higher energy false vacuum nucleates in an otherwise empty space-time, in the true, lowest energy vacuum state. This nucleation process is essentially the inverse of that discussed previously. If the nucleated bubble is sufficiently large, it will expand at an incredibly rapid pace like the inflationary epoch that took place early in the history of our universe. As the bubble expands, it eventually becomes causally disconnected from the original space-time of our universe. In this sense, the newly created bubble becomes a new and separate universe which can be considered as a "child universe."

This creation of a new universe is rather counterintuitive. How is an entire universe produced from nothing? Where or what does it expand into? The key idea is that the newly created bubble of new space-time is not expanding into anything, but rather the space-time itself is expanding. The new universe thus creates its own space and does not "use up" any of our universe.

To illustrate this concept, we can use an embedding diagram, a visualization technique often employed by physicists studying general relativity (as discussed in the previous chapter). We envision a two-dimensional version of the universe as a sheet of rubber, which stretches as the universe expands. The departure of the rubber sheet from a purely flat geometry represents the curvature of space-time, or equivalently, the degree to which space-time departs from being purely Euclidean or flat. This two-dimensional universe is thus embedded in a three-dimensional space that we can understand. The actual universe has a three-dimensional space with an additional time dimension; since we cannot readily visualize so many dimensions, we use this simple two-dimensional universe as a model.

Using this technique, we can visualize the creation of a new "child universe" as the production of an indentation or bubble in the rubber sheet, as shown in the figure opposite. As the newly created universe expands, it rapidly becomes a huge sphere that is connected to the old flat universe by a narrow tube. This highly curved narrow connection to the old universe is an example of a relativistic wormhole, a type of bridge that connects different parts of space-time. In this case, the wormhole smoothly connects the newly formed child universe to the flat space-time of the old universe. As the new universe expands, however, the wormhole itself shrinks and eventually evaporates away. When the wormhole disappears, the old universe no longer has any causal connection to the child universe, which then becomes truly separated.

This newly created universe appears quite different to observers living inside the bubble and those located on the outside. Observers living within the bubble see their local universe in a state of exponential expansion, much like the inflationary phase of our universe in its earliest moments of existence. On the other hand, observers situated outside the bubble, those left behind in the empty space-time background of our present universe, see the newly created universe as a collapsing black hole that quickly becomes causally disconnected from our universe. Because of this causal disconnection, these child universes cannot affect the future evolution of our universe.

Yet these child universes can receive information from our universe, at least in principle. Before the newly created universe grows out of causal contact with our own space-time, the two universes are connected through a relativistic wormhole, which can provide a conduit for information transfer and

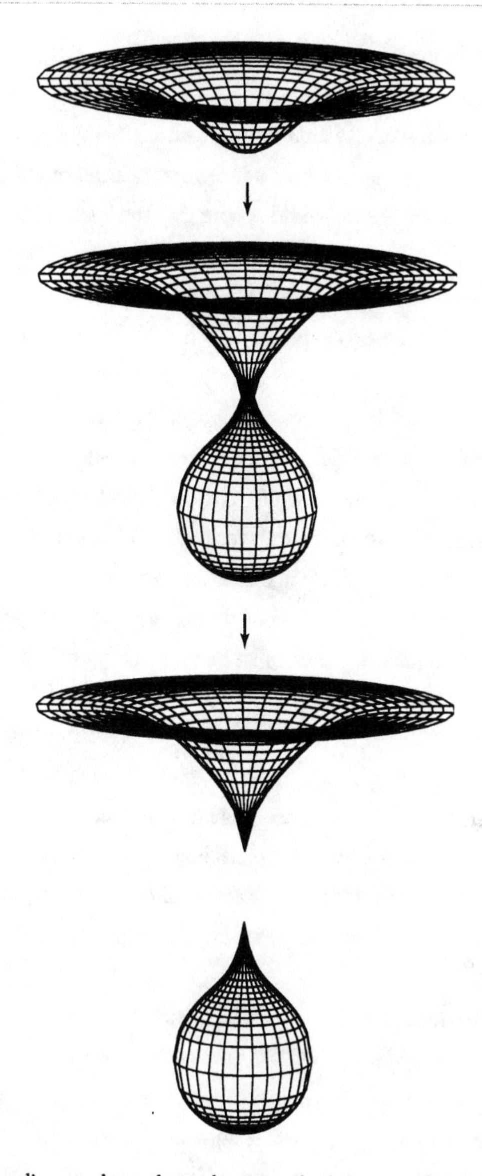

This embedding diagram shows the nucleation and subsequent development of a child universe. The top drawing depicts the nucleation event and hence the initial production of the child universe. In the middle drawing, the newly created universe is growing rapidly, but is still connected to the old universe (the flat region) by a narrow tube representing a relativistic wormhole. In the bottom drawing, the wormhole has evaporated and the child universe has detached itself from the original space-time from which it was born.

perhaps even the transfer of matter. The implications of this possibility are profound. Faced with inevitable death and extinction in our present universe, a sufficiently advanced civilization could create another universe as an avenue of escape. If they could predetermine the properties and physical laws of the new universe, the civilization could ensure that the infant cosmos would be fully conducive to the development of life. Information from our old and dying universe could be transferred into the new universe through the connecting wormhole, before it evaporates. Such information could include blueprints for life or whatever treasures of civilization were deemed important by the society in question. It might even be possible for matter itself to be transferred through the wormhole. Perhaps sufficiently intelligent beings could build a universe and literally jump into it before the gateway closes.

This concept of child universes is related to the idea of black hole singularities being gateways to other universes, as discussed in Chapter 4. In the previous case, we were implicitly considering large black holes that live for a long time. The singularity, which must be contained somewhere within the event horizon of the black hole, provides a long-term connection to the other universe, although the connection is essentially inaccessible. For a nucleation event leading to a child universe, however, the collapsing black hole (seen by observers in our universe) is rather small in mass and relatively short-lived. Viewed from the outside, this tiny black hole appears to radiate itself away and ultimately disappear through the action of Hawking evaporation. In this case, the gateway, in addition to being effectively inaccessible, is also highly transient.

The nucleation of a child universe can also take place at the singularity of an already-established black hole. As in the other cases discussed above, the newly created universe grows quickly out of causal contact. Since this entire episode takes place within the event horizon of the original black hole, the child universe appears to have no connection whatsoever to our universe and would seem to hold little interest. Surprisingly, however, the nucleation of a child universe does have a characteristic signature that can be measured in our universe, outside the event horizon. The nucleation event partially suppresses the Hawking radiation emitted by the black hole. This drop in the radiation output of the black hole can be measured. Once the child universe grows out of causal contact, the usual Hawking evaporation signature is resumed. Once again, quantum effects allow some type of information to be transmitted through the event horizon, even though such clandestine transport is classically forbidden.

Just as cosmic phase transitions can, in principle, be triggered by outside influences, the production of a child universe can also be induced. The technology required for this type of planned parenthood is not currently available, but physicists have already seriously discussed the theoretical possibility of producing a new universe within a laboratory. In the future, the production of new universes could even become routine and commonplace. Once we recognize this possibility, an important philosophical question immediately springs to mind: Was our universe planned or induced to exist through the directed action of some other beings?

The concepts of purposefully setting off a cosmological phase transition, or creating a new universe in a laboratory, are good examples of how we, or other sentient beings, can directly affect the evolution of the universe as a whole. These feats of cosmic engineering are still safely beyond our current technological capabilities, but quite within the range of possibility. In the past few hundred years, humans on Earth have been able, for better or worse, to seize some measure of control of the planet. We can now directly affect the environment, the populations and species of flora and fauna, and even the climate of our home world. Such control was utterly inconceivable only a few centuries ago, and that time span is minuscule compared to the vast expanses of time yet to come. In the future, it is not only possible, but perhaps even likely, that intelligent life can play a significant role in setting the long-term course of evolution, for our universe and others.

CONCLUSION

A GUIDE TO THE IMPOSSIBLE, THE IMPROBABLE, AND THE MARVELOUS.

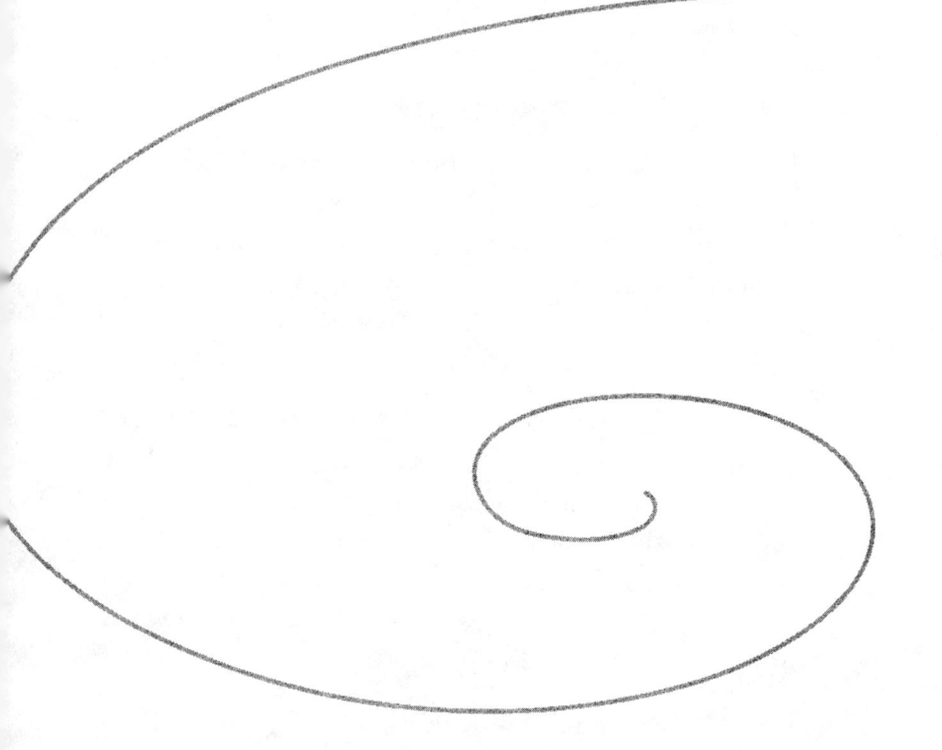

In a lonely garret not far from the British Museum:

Cornelius grabbed a clean sheet of paper, spun it through the roller, and began typing. The starting point for his saga was the big bang itself, when the cosmos embarked upon its ever expanding journey towards the future. After a short burst of inflation, the universe cascaded through a series of phase transitions and generated an excess of matter over antimatter. Throughout this Primordial Era, the universe had no cosmic structures of any kind.

After a million years and many reams of paper, Cornelius reached the Stelliferous Era, a span of time when stars vigorously churn through their life cycles and generate energy from nuclear reactions. This bright chapter draws to a close when galaxies run out of hydrogen gas, star formation ceases, and the longest-lived red dwarfs slowly fade away.

Typing persistently, Cornelius and his story enter the Degenerate Era, with its brown dwarfs, white dwarfs, neutron stars, and black holes. In the midst of this frozen desolation, dark matter slowly collects within dead stars and annihilates into radiation that powers the cosmos. Proton decay ushers in the end of this chapter as the mass-energy of the degenerate remnants slowly leaks away and carbon-based life reaches a definitive extinction.

As the weary author continues his task, the only remaining characters are black holes. But black holes cannot live forever. Shining ever so faintly, these dark objects evaporate through a slow quantum mechanical process. With no other energy source, the universe draws its power from this meager radiative output. After the largest black holes have evaporated, the awkward twilight of the Black Hole Era gives way to a deeper blackness.

Beginning his final chapter, Cornelius runs out of paper, but not out of time. The universe contains no stellar objects, only the leftover waste products from previous cosmic dramas. In this cold, dim, and far distant Dark Era, activity within the cosmos slows down markedly. The extraordinarily low levels of energy are matched with enormous expanses of time. After its fiery youth and energetic middle age, the universe now slowly shuffles into the darkness.

As the universe grows older, it continually changes its character. In each phase of its future evolution, the universe supports an astonishing variety of comple x physical processes and other interesting behavior. Our biography of the universe, from its explosive beginning to its long and gradual slide into the dark, is based on our current understanding of the laws of physics and the wonders of astrophysics. With the detailed and comprehensive state of today's science, this story represents the most probable vision of the future that we can construct.

MIND-BENDINGLY LARGE NUMBERS

As we discuss the wide range of exotic possible behavior in the future of our universe, one might get the idea that anything can happen. But this is simply not the case. In spite of the richness of physical possibilities, only a tiny fraction of the events that could occur will actually come about.

First and foremost, the laws of physics place great constraints on any allowed behavior. The total energy must be conserved. Electric charge must be conserved. A general guiding concept is the second law of thermodynamics, which formally states that the total entropy of a physical system must increase. Loosely speaking, this law implies that systems must evolve toward states of increasing disorder. In a practical sense, the second law enforces the idea that heat flows from hot objects to colder objects, rather than the other way around.

But even within the processes that are allowed by the laws of physics, many events that could take place in principle simply do not ever occur. One general reason is that they simply take too long and other processes happen first and supersede them. A good example of this trend is the process of cold fusion. As we discussed in conjunction with nuclear reactions in stellar interiors, the iron nucleus is the most stable of all possible nuclei. A collection of smaller nuclei, like hydrogen and helium, would give up energy if they could be assembled into iron. On the other end of the periodic table, larger nuclei like uranium would also give up energy if they could be broken apart and the pieces were arranged into iron nuclei. Iron represents the lowest energy state available to nuclei. Nuclei want to be in the form of iron, but energy barriers inhibit this transformation from readily taking place under most circumstances. Either high temperatures or long waiting times are generally required to overcome these energy barriers.

Consider a large piece of solid matter, like a rock or perhaps a planet. The structure of the solid is frozen into place through ordinary electromagnetic forces, like those involved in chemical bonding. Instead of retaining its original nuclear composition, the material could in principle rearrange itself so that all of its atomic nuclei are made of iron. In order for this restructuring of matter to occur, the nuclei must overcome the electrical forces that hold the matter in place and the repulsive electrical forces that the nuclei exert on each other. These electrical forces create a strong energy barrier, much like the one depicted in the figure on page 171. Because of this barrier, the nuclei must rearrange themselves through a quantum mechanical tunneling process (once the nuclei penetrate the barrier, the attractive strong force instigates fusion). In this manner, the lump of material would exhibit nuclear activity. Given enough time, the entire rock or planet would become pure iron.

How long would this nuclear rearrangement take? Approximately fifteen hundred cosmological decades are required for this type of nuclear activity to transform the nuclei within rocks into iron. If this nuclear process could take place, excess energy would be emitted because the iron nuclei correspond to a lower energy state. But this cold nuclear fusion process will never proceed to completion. It will never even really get started. All of the protons in nuclei decay into smaller particles long before the nuclei transform themselves into iron. Even the longest possible proton lifetime is less than two hundred cosmological decades, vastly shorter than the enormous time interval necessary for cold fusion. Stated another way, nuclei disintegrate before they have a chance to fuse into iron.

Another physical process that takes too long to be cosmologically important is the tunneling of degenerate stars into black holes. Because black holes are the lowest energy states available to stars, a degenerate object like a white dwarf has more energy than a black hole of the same mass. The white dwarf would thus release energy if it could spontaneously transform itself into a black hole. This transformation does not normally occur because of the energy barrier provided by degeneracy pressure, which holds up the white dwarf.

In spite of the energy barrier, the white dwarf could transform itself into a black hole through quantum mechanical tunneling. Because of the uncertainty principle, all of the particles (10^{57} or so) that make up the white dwarf could find themselves within such a small space that they form a black hole. The time required for such a fortuitous event to take place, however, is extraordinarily long, about 10^{76} cosmological decades. One cannot overemphasize the truly enormous nature of 10^{76} cosmological decades, which is the same as $10^{10^{76}}$ years. The number for this vast time span, in years, would be written as a 1 followed by 10^{76} zeros. We could not even begin to write down this number in a book; it would have about one zero for every proton in the observable universe of today, give or take a couple orders of magnitude. Needless to say, protons will decay and white dwarfs will be long gone before the universe reaches the $10^{76\text{th}}$ cosmological decade.

WHAT REALLY HAPPENS IN LONG-TERM EXPANSION

While many events are effectively impossible, a wide range of possibilities remain. The broadest categories of future cosmological behavior are based on whether the universe is open, flat, or closed. An open or flat universe will expand forever, whereas a closed universe recollapses after some given time which depends on the initial state of the universe. Considering more speculative possibilities, however, we find that the future evolution of the universe can be much more complicated than this simple classification scheme suggests.

The basic problem is that we can only make physically meaningful measurements, and hence definitive conclusions, about the local portion of the universe, the part within our cosmological horizon of today. We can measure the total density of the universe inside this local region, which is about twenty billion light years in diameter. Unfortunately, however, measurements of the density within this local volume do not necessarily determine the long-term fate of the universe as a whole, which can be vastly larger.

Suppose, for example, that we were to measure the cosmological density to be greater than that required to close the universe. We would thus tentatively conclude that our universe must recollapse in the future. The universe would apparently be slated to go through an accelerating sequence of natural disasters leading to the big crunch, as outlined in the next section. But that is not the entire story. Our local portion of the universe, the part we observe to be closed in this scenario of apparent armageddon, could be embedded within a much larger region with much lower density. In this case, only part of the total universe would collapse. The remainder, encompassing perhaps most of the universe, could keep expanding without bound.

One might argue that this added complication does little good—our own part of the universe would still be destined to collapse. Death and destruction would not be avoided locally. This glimpse at the larger picture does, however, change our perspective substantially. If the larger universe as a whole still survives, the death of our local region is not as tragic. While the destruction of one city on Earth, say due to an earthquake, is undeniably terrible, it is not nearly as bad as the total annihilation of the planet. In the same manner, the loss of one small part of the total universe is not as devastating as the loss of the whole. Complex physical, chemical, and biological processes could still play themselves out into the far future, somewhere in the universe. The destruction of our local universe would be just another catastrophe in the continuing series of astrophysical disasters the future might bring: the death of our Sun, the end of life on Earth, the evaporation and dispersal of the galaxy, the decay of protons and hence the destruction of all ordinary matter, the sublimation of black holes, and so on.

The survival of the larger universe allows for the possibility of escape, either actual long-distance travel or a vicarious escape through sending information via light signals. The escape route might be complicated or even nonexistent—it depends on how the closed region of our local space-time matches onto the larger region of the universe. The fact that life can continue elsewhere, however, acts to keep hope alive.

If our local region recollapses, we might not have time for all of the astronomical events outlined in this book to take place locally. These processes would, however, eventually occur elsewhere in the universe, far away from here. How much time we have until the local universe recollapses depends on the local density. Although current astronomical measurements suggest that the density is low enough that our local universe will not recollapse at all, additional matter could be lurking unseen in the darkness. The largest possible allowed value for the local density is about twice that needed for closure. Even with this maximum density, the universe cannot begin to collapse for at least twenty billion years. This time frame would give us at least another fifty billion years before the local version of the big crunch.

The opposite set of circumstances can also arise. Our local portion of the universe could exhibit a relatively low density and hence apparently be set to live forever. However, this local patch of space-time could be embedded within a much larger region with higher density. In this case, when our local cosmological horizon grows big enough to encompass the larger region of higher density, our local universe would become part of a greater universe that is destined to recollapse.

This scenario of destruction requires that our local universe have nearly a flat cosmological geometry so that the expansion rate continues to decline steadily. The almost flat geometry allows larger and larger regions of the meta-scale universe (the big picture universe) to affect local events. This larger encompassing region must be just barely dense enough to eventually recollapse. It must live long enough (that is, not recollapse too soon) so that our cosmological horizon can grow to the requisite large size scale.

If these ideas are actualized in the cosmos, then our local universe is not "the same" as the much larger region of the universe that engulfs it. The cosmological principle would thus be manifestly violated on sufficiently large sizes—the universe would not be the same everywhere in space (homogeneous) and not necessarily the same in all directions (isotropic). This potentiality does not invalidate our use of the cosmological principle for studies of past history (as in big bang theory) because the universe is apparently homogeneous and isotropic within our local region of space-time, which is now about ten billion light years in radius. These potential departures from homogeneity and isotropy are restricted to larger sizes and hence can only manifest themselves in the future.

Somewhat surprisingly, we can put constraints on the nature of the larger portion of the universe currently outside our cosmological horizon. The cosmic background radiation is measured to be extraordinarily smooth. But large differences in density of the universe, even if the differences are outside the horizon, would create ripples in this smooth radiation background. The absence of large ripples thus implies that any putative large density variations must be far away. If the large density fluctuations are far away, then our local region of the universe can live for a rather long time before encountering them. The soonest possible time for a large density variation to exert its effects on our portion of the universe is about 17 cosmological decades. But such a universe-altering event is likely to take far longer to occur. Most versions of the inflationary universe theory allow the universe to remain smooth and nearly flat for hundreds and even thousands of cosmological decades.

THE BIG CRUNCH

If the universe (or a portion of it) is closed, then gravity would win its battle with the expansion and usher in an inevitable collapse. Such a recollapsing universe would end its life in a fiery denouement known as *the big crunch*. Many of the vicissitudes that mark the time-line of a collapsing universe were first discussed by Sir Martin Rees, now the Astronomer Royal of England. No shortage of catastrophes occurs if the universe plunges into this grand finale.

Although the universe is likely to expand forever, we know with relative certainty that the universe contains less than twice the critical density. With this upper limit, the *smallest* possible amount of time left before the universe could collapse into a big crunch is about fifty billion years. Doomsday is still a long way off by any human standard of time, and one should probably continue paying the rent.

Suppose that the universe does recollapse after reaching a maximum size in twenty billion years. At this point, the universe would be about twice as large as it is today. The background radiation temperature would be about 1.4 degrees kelvin, half its present value. After the universe cools to this minimum temperature, the subsequent collapse would heat up the universe as it plummets toward a big crunch. Along the way, all of the structures produced by the universe—clusters, galaxies, stars, planets, and even the elements themselves—would be destroyed during the collapse.

Approximately twenty billion years after the universe begins to recollapse, it would return to the size and density of today's universe. During the intervening forty billion years, the universe forges ahead with a roughly constant large scale appearance. Stars would continue to form, evolve, and die. Small fuel-conserving stars, like our close neighbor Proxima Centauri, don't have enough time to evolve in any significant way. Some galaxies collide and merge within their parent clusters, but most persist in largely unaltered condition. An isolated galaxy requires vastly more than forty billion years to modify its dynamical structure. In a mounting inversion of Hubble's law, some galaxies would approach our galaxy rather than recede from it. Only this curious trend of blueshifts would reveal to astronomers a glimpse of the impending catastrophe.

Distinct galaxy clusters, scattered throughout the vastness of space and loosely organized into clumps and filaments, would remain intact until the universe contracts to a size five times smaller than at present. At this hypothetical future juncture, clusters of galaxies merge. Within the present universe, galaxy clusters fill only about 1 percent of the volume. When the universe contracts to a fifth of its current size, however, the clusters effectively fill all of space. The universe would thus become one giant cluster of galaxies, but the galaxies themselves would retain their individuality during this epoch.

As the collapse continues, the universe would soon become one hundred times smaller than today. When this milestone is reached the average density of the universe would be the same as the average density within a galaxy. The galaxies would overlap and individual stars would no longer belong to any particular galaxy. The whole universe then becomes one giant galaxy filled with stars. The background temperature of the universe, provided by the cosmic background radiation, grows to 274 degrees kelvin, near the melting point of ice. With the escalating compression of events beyond this epoch, it becomes more convenient to speak in terms of the other end of time, the time remaining before the big crunch. When the temperature of the universe reaches the melting point of ice, ten million years of future history are left.

Until this point, life on Earthlike planets would continue relatively unaffected by the cosmic evolution taking place in the background. Indeed, the warmth of the sky would eventually melt frozen Pluto-like objects, which drift through the outer reaches of every solar system, and provide a last fleeting chance for life to flower throughout the universe. This relatively brief and final spring would draw to a close as the temperature of the background radiation continues to escalate. As liquid water disappears, mass extinctions take place more or less simultaneously throughout the universe. The oceans boil away, and the night sky grows brighter than the daytime sky seen on Earth today. With only six million years remaining before the final crunch, any surviving life forms must either burrow deep within planets, or evolve elaborate and efficient cooling mechanisms.

After the clusters, and then the galaxies, have been effectively destroyed, the stars find themselves next in the line of fire. If nothing else happened, the stars would eventually collide and tear each other apart in the face of continuing and crushing collapse. This particular violent fate is circumvented, however, because the stars would be destroyed in a more gradual manner long before the universe grows dense enough for stellar collisions. When the temperature of the relentlessly contracting background radiation exceeds the typical surface temperature of a star, 4000 to 6000 degrees kelvin, the radiation field can drastically alter the structure of the stars. Although nuclear reactions continue unhindered within stellar cores, the stellar surfaces are evaporated away by the intense external radiation field. The background of radiation thus serves as the ultimate cause of stellar destruction.

When the stars start to evaporate, the universe is about two thousand times smaller than it is today. During this turbulent epoch, the night sky would appear as bright as the surface of the Sun. The shortness of the time remaining would be hard to ignore; the intense radiation burns through any uncertainty that less than a million years remain until the end. Any astronomers with the technological acumen to survive into this era might recall with resigned amusement that their seething cauldron of a universe—stars embedded within a sky as bright as the Sun—is a revisitation of Olbers' paradox for an infinitely old and static universe (see the box on page 21).

Any stellar cores, or brown dwarfs, that survive into the epoch of evaporation are unceremoniously blown apart. When the background radiation reaches ten million degrees kelvin, comparable to the present conditions in the central regions of stars, any remaining nuclear fuel can be ignited in an increasingly explosive and dramatic fashion. Stellar objects that manage to survive evaporation thus contribute to the general atmosphere of armageddon by becoming futuristic fusion bombs.

The fate of the planets in a collapsing universe is similar to that of the stars. Gaseous giant planets, such as Jupiter and Saturn, evaporate much more easily than the stars and leave behind their central cores, which are indistinguishable from terrestrial planets. Any liquid water would have long since evaporated from the planetary surfaces and the atmospheres would soon follow. Only barren and sterile wastelands are left behind. The rocky surfaces melt and the layers of liquid rock gradually grow deeper, eventually leaving the planets completely molten. Gravity holds the dying molten remnants together as they build up crushing silicate atmospheres, which in turn bleed into space. The evaporating planets vanish without a trace into the blinding heat.

As planets depart the scene, the atoms of interstellar space begin breaking apart into their constituent nuclei and electrons. The background radiation field is so intense that the photons (particles of light) have enough energy to liberate the electrons. As a result, atoms cease to exist during the final few hundred thousand years, and matter is broken down into charged particles. The background radiation field interacts strongly with these charged particles and hence the matter and the radiation are closely coupled. Cosmic background photons, which have traveled unobstructed for nearly sixty billion years since the epoch of recombination, meet their surface of "next" scattering.

A Rubicon is crossed when the universe shrinks to a ten-thousandth of its current size. At this point, the density of radiation surpasses the density of matter, a situation not seen since shortly after the big bang. The universe once again becomes radiation dominated. Because matter and radiation behave differently as they are compressed, the subsequent collapse is slightly altered as the universe passes through this transition. Only ten thousand years are left.

With three minutes remaining before the final crunch, atomic nuclei begin to break apart. This particular brand of mayhem continues until only one second remains, when all free nuclei have been destroyed. This epoch of antinucleosynthesis is quite different from the frenzy of nucleosynthesis that occurred during the first few minutes of the Primordial Era. During the first few minutes, only the lightest elements were produced, mostly hydrogen, helium, and a touch of lithium. During these final three minutes, however, heavy nuclei of all varieties are present. The iron nuclei are the most tightly bound together, and they require the most energy per particle to break apart. A collapsing universe, however, produces increasingly higher temperatures and energies; even the iron nuclei must eventually succumb in this ferociously disruptive environment. For the final second of the universe's life, none of the elements remains. The protons and neutrons are free again, just as they were during the first second of cosmic history.

If any life remains in the universe at this epoch, the destruction of nuclei represents a point of no return. Nothing even remotely resembling the carbon-based life that we have on Earth would remain after this time. No carbon remains in the universe. Any organism that can transcend the dissociation of nuclei must be of a truly exotic variety. Perhaps beings based on the strong force could bear witness to the last second.

The final second is much like a film of the big bang shown backwards. After the destruction of nuclei, when the universe is only a microsecond from its end, the protons and neutrons themselves break apart, and the universe becomes a sea of free quarks. As the collapse continues, the universe grows hotter and denser, and the laws of physics seem to change. When the universe reaches a temperature of about 10^{15} degrees kelvin, the weak nuclear force and the electromagnetic force unite to form the electroweak force. This event represents a kind of cosmological phase transition, roughly analogous to the melting of ice into water. As we go to higher energies, probing ever closer to the end of time, we become further removed from direct experimental confirmation and the discussion necessarily becomes more speculative. We continue nonetheless. After all, the universe still has 10^{-11} seconds of history left.

The next important transition occurs when the strong force unites with the electroweak force. This event, known as *grand unification,* unifies three of the four fundamental forces of nature: the strong, weak, and electromagnetic forces. The unification takes place at an incredibly high temperature, 10^{28} degrees kelvin, with only 10^{-37} seconds remaining.

The last major event that we can put on the calendar is the unification of gravity with the other three forces. This milestone takes place when the collapsing universe reaches a temperature of about 10^{32} degrees kelvin and only 10^{-43} seconds remain before the big crunch. This temperature or energy scale is generally known as the *Planck scale.* Unfortunately, scientists do not have a self-consistent theory of physics at this energy scale, which marks the consolidation of all four fundamental forces of nature. When this unification of forces occurs during the recollapse, our current understanding of physical law breaks down. We don't know what happens next.

THE FINE-TUNING OF OUR UNIVERSE

Having looked at impossible events, and improbable events, let's now consider the most extraordinary event that did take place—the ascent of life. Our universe is rather convenient for life as we know it. In fact, all four windows of astrophysics play a vital role in its development. Planets, our smallest window of astronomy, provide the home for life. They provide the petri dishes in which life can arise, evolve, and develop. Stars are also obviously important, as they provide the energy source that drives biological evolution. Stars also play a second fundamental role, as the alchemists that produce the elements heavier than helium—the carbon, oxygen, calcium, and other nuclei that make up life forms.

Although less obvious, the galaxies are also extremely important. Without the binding influence of galaxies, the heavy elements produced by stars would spread out over the universe. These heavy elements are the essential building blocks of both planets and life-forms. The galaxies, with their large masses and strong gravitational attraction, keep together the chemically enriched gas left over from stellar death. This previously processed gas is then incorporated into future generations of stars, planets, and people. The gravitational attraction of galaxies thus ensures that heavy elements are readily available for successive generations of stars, and for the production of rocky planets like our Earth.

On the largest size scale, the universe itself must have the right properties to allow the development of life. Although we do not have anything remotely approaching a complete understanding of life and its evolution, one basic requirement is relatively certain: it takes a long time. The ascent of mankind required about four billion years on our particular planet and we wager that at least one billion years is required for intelligent life to arise in any case. The universe as a whole must thus live for billions of years to allow the development of life, at least for biology even vaguely resembling ours.

The properties of our universe as a whole also conspire to provide a chemical environment that is conducive to the evolution of life. Although heavier elements like carbon and oxygen are synthesized within stars, hydrogen itself is also a vital component. It makes up two out of three atoms in water, H_2O, an essential ingredient for life on our planet. In considering the vast ensemble of possible universes and their possible properties, we note that primordial nucleosynthesis could have processed all of the hydrogen into helium and even heavier elements. Alternately, the universe could have expanded so rapidly that protons and electrons never got together to make hydrogen atoms. In either case, the universe could have ended up with no hydrogen atoms to make water molecules and hence no ordinary life.

With these considerations, the fact emerges that our universe does indeed have the proper special features to allow for our existence. Given the laws of physics, as they are determined by the values of the physical constants, the strengths of the fundamental forces, and the masses of elementary particles, our universe naturally produces galaxies, stars, planets, and life.

With only a slightly different version of the physical laws, our universe could have been completely uninhabitable and astronomically impoverished.

Let's illustrate the required fine-tuning of our universe a bit more. Galaxies, one of the astrophysical entities required for life, are produced as gravity wins its battle with the expansion of the universe and instigates the collapse of local regions. If the gravitational force was much weaker, or the cosmological expansion rate was much faster, then no galaxies would have formed within the current age of the cosmos. The universe would continue to spread itself out, but would contain no gravitationally bound structures, at least not by this time in cosmic history. On the other hand, if the gravitational force was much stronger or the expansion rate was much slower, then the entire universe would recollapse into a big crunch long before galaxies even begin to form. In either case, no life would evolve in our present universe. The interesting case of a universe filled with galaxies and other large cosmic structures thus requires a reasonably delicate compromise between the strength of gravity and the expansion rate. And our universe has realized just such a compromise.

In the case of stars, the required fine-tuning of physical theory is even more severe. Nuclear fusion reactions in stars fill two vital roles required for the evolution of life: the generation of energy and the production of heavy elements such as carbon and oxygen. In order for stars to play their part, they must live for a long time, achieve sufficiently high central burning temperatures, and be relatively common. The universe must be endowed with a wide range of special properties for all of these pieces of the puzzle to come together.

Perhaps the most readily explained example arises from nuclear physics. Nuclear fusion reactions and nuclear structure depend on the magnitude of the strong nuclear force. Atomic nuclei exist as bound structures because the strong force is sufficiently powerful to hold protons together, even as the electrical repulsion of positively charged protons acts to disperse the nucleus. If the strong force had been slightly weaker, then heavy nuclei would simply not exist. The universe would contain no carbon and hence no carbon-based lifeforms. On the other hand, if the strong nuclear force had been even stronger, then two protons could be bound together into pairs called diprotons. In this case, the strong force would be so strong that all of the protons in the universe would become bound into diprotons, or even larger nuclear structures, and no ordinary hydrogen would remain. With no hydrogen, the universe would contain no water and hence no familiar varieties of life-forms. Fortunately, for us, our universe is graced with the proper type of strong nuclear force to allow for hydrogen, water, carbon, and other necessary ingredients for life.

In a similar vein, if the weak nuclear force had a much different strength, stellar evolution would be greatly affected. If the weak force was much stronger, comparable to the strong force for example, then nuclear reactions within stars would proceed at much faster rates and stellar lifetimes would be drastically reduced. A name change for the weak force would also be in order. The universe has some leeway on this issue because of the range of stellar masses—smaller stars live longer and could be used to drive biological evolution in place of our Sun. However, degeneracy pressure (from quantum mechanics) prevents stars from burning hydrogen if they become too small in mass. So even the longest-lived stars would have their lifetimes seriously shortened. Once the maximum stellar lifetime drops below the billion year mark, the development of life is clearly in trouble. The actual magnitude of the weak force, millions of times smaller than the strong force, allows the Sun to burn its hydrogen in the slow and leisurely manner required for the evolution of life on Earth.

Next we must consider planets, the smallest astrophysical objects necessary for life. The formation of planets requires the universe to produce heavy elements and hence requires the same nuclear constraints outlined already. In addition, planets require that the background temperature of the universe is sufficiently cool for solids to condense. If our universe was only six times smaller than it is today, and hence one thousand times hotter, then interstellar dust grains would evaporate and no raw material would be available for forming rocky planets. Even the formation of giant planets would be highly suppressed in this hot and hypothetical universe. Fortunately, our universe is cool enough to allow the planets to arise.

Another consideration is the long-term stability of a solar system once it has formed. In our galaxy today, stellar interactions and close encounters are both rare and weak because the density of stars is so low. If our galaxy contained the same number of stars but happened to be one hundred times smaller, the enhanced stellar density would lead to a rather high probability of another star entering our solar system and disrupting the orbits of the planets. Such a cosmic collision could alter Earth's orbit and render our planet uninhabitable, or could eject Earth from the solar system altogether. In either case, such a cataclysmic event would mark the end of life. Luckily, in our galaxy, the expected time for our solar system to experience a course-altering collision is much longer than the time required for life to develop.

We see that a long-lived universe that contains galaxies, stars, and planets requires a rather special suite of values for the fundamental constants that determine the strengths of the basic forces. This required fine-tuning thus sets up a basic question: *Why does our universe happen to have these particular properties that ultimately give rise to life?* It is a truly remarkable coincidence that physical laws are just right to allow for our existence. It seems as if the universe somehow knew that we were coming. Of course, if the conditions were otherwise, we would simply not be here to contemplate this issue. But the nagging question remains—why?

Understanding *why* the physical laws are the way they are places us at the edge of development of present-day science. Preliminary explanations have been put forward, but the question remains open. While 20th century science has developed a good working understanding of *what* our physical laws are, we can hope that the science of this coming century will provide us with an understanding of *why* physical laws have the form that they do. Scientific hints along these lines are already starting to develop, as we will now consider.

ETERNAL COMPLEXITY

The seeming coincidence that the universe has the requisite special properties that allow for life suddenly seems much less miraculous if we adopt the point of view that our universe, the region of space-time that we are connected to, is but one of countless other universes. In other words, our universe is but one small part of a *multiverse,* a large ensemble of universes, each with its own variations of physical law. In this case, the entire collection of universes would fully sample the many different possible variations of the laws of physics.

However, only those particular universes with the proper versions of physical law would develop life. The fact that we happen to live in a universe with the right properties for life then becomes obvious.

We must clarify the difference between "other universes" and "other parts" of our own universe. The large scale geometry of space-time could be very complicated. We now live in a smooth patch of the universe that is about twenty billion light-years in diameter. This region represents the portion of space that can causally affect us, so far in time. As the universe ages into the future, the region of space-time that affects us will grow larger. In this sense, our universe will contain more of space-time as it gets older. However, there can be other regions of space-time that will *never* be in causal contact with our part of the universe, no matter how long we wait, no matter how old our universe becomes. These other regions grow and evolve completely separately from the physical events occurring within our universe. Such regions belong to other universes.

Once we allow for the possibility of other universes, the set of coincidences presented by our particular universe seems much more palatable. But does the concept of the existence of other universes actually make any sense? Can multiple universes be naturally accommodated within a theory like the big bang, or through sensible extensions? Surprisingly, the answer is a resounding yes.

Andrei Linde, an eminent Russian cosmologist now at Stanford, has introduced the concept of *eternal inflation*. Roughly speaking, this theoretical idea means that at all times, some region of space-time, somewhere in the multiverse, is undergoing an inflationary phase of expansion. In this scenario, the space-time foam is always launching new universes into existence through the inflationary mechanism (as discussed in Chapter 1). Some fraction of these inflating regions evolve to become interesting universes like our own local patch of space-time. They have physical laws that drive the formation of galaxies, stars, and planets. Some of these regions may even develop intelligent life.

This idea both makes sense physically and holds a great deal of intrinsic appeal. Even if our universe, our own local region of space-time, is slated to die a slow and tortuous death, other universes will always be around. Something will always be in existence. Viewed from the grander perspective encompassing the whole ensemble of universes, the multiverse can be considered truly eternal.

This picture of cosmic evolution elegantly sidesteps one of the most vexing questions arising from 20th century cosmology: *If the universe started as a big bang that took place only ten billion years ago, what happened before the big bang?* This perplexing question of "before-the-beginning" represents the boundary between science and philosophy, between physics and metaphysics. We can extrapolate physical law back to when the universe was only 10^{-43} seconds old, albeit with increasing uncertainty as we go back in time, but even earlier epochs are inaccessible to current scientific methods. However, science is always pushing its boundaries outwards and is starting to progress in this arena. Within the greater context afforded by the concept of the multiverse and eternal inflation, we can actually formulate an answer: Before the big bang, there existed (and still exists!) a foamy region of high-energy space-time. Out of this cosmic froth, some ten billion years ago, our own particular universe was launched into existence and continues to evolve today. Other universes are continually being launched in a similar manner and this process can proceed indefinitely. Admittedly, this answer remains a little vague, and perhaps somewhat unsatisfying. Nevertheless, physics has advanced to a state in which we can begin to address this long-standing question.

With the concept of the multiverse in place, the next battle of the Copernican revolution is thrust upon us. Just as our planet has no special status within our solar system, and our solar system has no special location within the universe, *our universe has no special status within the vast cosmic melange of universes that comprise the multiverse.*

A DARWINIAN VIEW OF UNIVERSES

The space-time of our universe becomes increasingly complicated as the universe grows older. In the beginning, immediately following the big bang, our universe was very smooth and homogeneous. These initial conditions were necessary in order for the universe to evolve into its present-day form. As the universe evolves, however, both galactic and stellar processes produce black holes, which puncture space-time with their interior singularities. Black holes thus produce what can be considered to be holes in space-time. In principle, these singularities can also provide connections to other universes. It is also possible that new universes, the child universes introduced in Chapter 5, can be nucleated at the singularity of a black hole. In this case, our universe would give birth to a new universe that is connected to ours through the black hole.

If we follow this chain of reasoning to its logical limit, an extremely interesting scenario arises for the evolution of universes within the multiverse. If universes can give birth to new universes, then the concepts of heredity, mutations, and even natural selection can arise in physical theory. This evolutionary picture has been championed by Lee Smolin, a physicist who is an expert on general relativity and quantum field theory.

Let's suppose that the singularities inside black holes can give rise to other universes, like the nucleation of child universes discussed in the previous chapter. As they evolve, these other universes generally grow causally disconnected from our own. These child universes are, however, tied to our universe through the singularity at the black hole center. Let's further suppose that the laws of physics in these new universes are similar to those in our universe, but not quite the same. In practice, this statement means that the physical constants, the strengths of the fundamental forces, and the particle masses have similar values, but are not identical. In other words, the child universe inherits a set of physical laws from its parent universe, but the laws can be slightly different, much like the mutations of genes during reproduction of Earthly biota. In this cosmological setting, the growth and behavior of the child universe will be similar to, but not exactly the same as, that of the original parent universe. This picture of heredity for universes is thus fully analogous to that of biological life-forms.

With both heredity and mutations, this ecosystem of universes attains the intriguing possibility of a Darwinian evolutionary scheme. The universes that are "successful" from this cosmological-Darwinian point of view are those that produce large numbers of black holes. Since black holes result from the formation and death of stars and galaxies, these successful universes must contain large numbers of stars and galaxies. Black holes also require a rather long time to form. Galaxies in our universe form on a billion-year time scale;

massive stars live and die on a shorter time scale of only millions of years. Any successful universe must be relatively long-lived in addition to having the proper values of the physical constants to allow for the formation of large numbers of stars and galaxies. With stars, galaxies, and a long life span, the universe has a fighting chance to develop life. In other words, successful universes automatically have nearly the right characteristics for the ascent of biological life-forms.

The evolution of the entire composite collection of universes proceeds in a manner analogous to biological evolution here on Earth. The successful universes produce large numbers of black holes and give birth to large numbers of child universes. These astronomical offspring inherit different kinds of physical laws, with slight variations from their parent universes. The mutations that lead to the production of more black holes also lead to the production of more offspring. As this ecosystem of universes progresses, the universes that are found in the greatest abundance, by far, are those which produce prodigious quantities of black holes, stars, and galaxies. These same universes have the best chances of producing life. Our universe, for whatever reason, has the proper characteristics to live for a long time and produce many stars and many galaxies—our own universe is successful according to this grand Darwinian scheme. Viewed from this enlarged perspective, our universe is neither unusual nor fine-tuned, but rather is common and hence should be expected. Although it remains speculative and controversial, this evolutionary picture provides an elegant and appealing explanation for why our universe has its observed properties.

PUSHING THE LIMITS OF TIME

In this cosmic biography, we have followed the universe from its brilliant, singular inception, through the warm familiar skies of today, across the bizarre frozen wastes, and as close to its ultimate dark demise as possible. As we attempt to peer ever deeper into the dim abyss, our powers of prediction necessarily falter. Our vicarious travels through cosmic time must therefore terminate, or at least become woefully incomplete, at some future epoch. In this book, we have constructed a time line which encompasses hundreds of

A Space-Time Capsule

Several times during this biography of the universe, we have faced the possibility of sending communication signals to other universes. If one could make a universe in a laboratory, for example, a coded signal could be imparted into it before it grows causally disconnected from our own. But if you could send such a message, what would you say?

One might wish to preserve the essential essence of our civilization: art, literature, and science. Every reader will have some idea of what parts of our culture should be preserved in this manner. While everyone will undoubtedly have a different opinion on this matter, we would be remiss if we did not offer some suggestion for archiving some portion of our culture. As an example, we offer an encapsulated version of science, or more precisely, physics and astronomy. The most basic messages to be passed on could include the following:

- Matter is composed of atoms, which are composed of smaller particles.
- Particles exhibit wavelike properties at small size scales.
- Nature is governed by four fundamental forces.
- The universe consists of a space-time that evolves.
- Our universe contains planets, stars, and galaxies.
- Physical systems evolve toward states of lower energy and increasing disorder.

These six points, whose universal role should now be clear, can be considered the jewels of our accomplishments in the physical sciences. They are perhaps the most important physical concepts discovered thus far by our civilization. If these concepts are the jewels, then surely the crown itself must be the scientific method. With the scientific method in place, all of these results follow automatically, given enough time and effort. If one could send but a single concept to another universe to represent the intellectual achievements of our culture, the scientific method would be a worthy ambassador.

cosmological decades. Undoubtedly some readers will feel that we have been far too arrogant in extending the story this far, whereas others might wonder how we could stop at a point which, when compared to eternity, is so cautiously close to the beginning.

Of one thing we can be sure. In its journey to the obscurity of the future, the universe displays a remarkable, entangled combination of transience and permanence. Although the universe itself will endure, virtually nothing within the future will resemble the present. The most consistent characteristic of our ever-evolving universe is change. And this universal process of incessant change demands an enlarged cosmological perspective, in other words, a profound shift in our view of the largest frames. Because the universe is continually changing, we must seize the present cosmological epoch, the present year, and even the present day. Each moment in the unfolding history of the cosmos represents a unique opportunity, a chance for greatness, an adventure to undertake. In keeping with the Copernican time principle, each future time period is rife with new possibilities.

It is not enough to make the passive assertion that events are inevitable, "to let determined things hold unbewailed their way." A passage often attributed to Huxley claims that "six monkeys, set to strum unintelligently on typewriters for millions of years, would be bound in time to write all the books in the British Museum." These imagined monkeys have long been indentured in the service of vague or defeatist thought, as a way to justify extraordinarily improbable events, or to implicitly minimize the great works of human accomplishment as being somehow drawn from a hat. After all, if something can happen, then it will happen, right?

Even our rudimentary understanding of the cosmic future reveals the sheer absurdity of this point of view. A simple calculation shows that the random monkeys require nearly a half a million cosmological decades, vastly more years than the number of protons in the universe, to produce a single particular book by chance.

The nature of the universe is scheduled to change completely, many times, before the random monkeys can even begin to complete their task. The monkeys must die of old age in less than one hundred years. The red giant Sun bakes the Earth, and hence the typewriters, into oblivion in five billion years.

The stars burn out completely, throughout the universe, in 14 cosmological decades, and the monkeys would no longer be able to see the keys. The galaxy loses its integrity by the 20th cosmological decade, and the monkeys stand a fair chance of being consumed by the black hole in the galactic center. Even the protons that make up the monkeys, and their work, are slated to decay before 40 cosmological decades have gone by, again, long before the Herculean task is even well under way. If the monkeys could endure this catastrophe, and continue their work by the feeble radiation emitted by black holes, their efforts would be compromised in the 100th cosmological decade as the last black holes make their explosive exits. Should the monkeys survive even longer, say to the 150th cosmological decade, they continue to face the ultimate hazard of a cosmological phase transition.

Although monkeys, typewriters, and typewritten manuscripts have been decimated many times over by the 150th cosmological decade, time itself certainly does not come to an end. In staring at the murkiness of the future, we are limited more by our lack of imagination, and perhaps by an inadequate physical understanding, than by a truly evaporated firmament of detail. The lower energy levels and apparent deficit of activity in store for the universe are more than compensated for by the increased amount of available time. We can put an optimistic face on an uncertain future. Although our comfortable world is destined to pass away, a wide variety of fascinating physical, astronomical, biological, and perhaps even intellectual events are waiting to unfold as our universe continues its voyage into the dark.

GLOSSARY

Active galactic nucleus. The central region of a galaxy with an energetic luminosity source, thought to be powered by a supermassive black hole.

Antimatter. Every type of particle has an associated antiparticle, another particle with the same mass but opposite charges. These antiparticles can annihilate with their particle partners. Antimatter is composed of these antiparticles.

Baryogenesis. The production of a net excess of baryons over antibaryons. In other words, the process that drives our universe to be made primarily of matter, rather than antimatter.

Baryon. A composite particle composed of three quarks, any three of the six possible quarks. Most baryons in our universe are protons and neutrons.

Baryonic matter. Ordinary matter made up of protons and neutrons, and perhaps other baryons.

Big bang. Explosive event at the beginning of the evolution of the universe. This event occurs at time $t = 0$ and represents a state of infinite density and temperature.

Big crunch. The final event at the end of the evolution of a recollapsing (closed) universe. This event represents a state of infinite density and temperature.

Blackbody. An object with a constant temperature that absorbs all radiation incident upon it; such an object emits light with a well-defined spectrum of radiation.

Black hole. A region of space-time in which the gravitational field is so strong that light cannot escape.

Black Hole Era. Time period in the future of the universe when black holes are the most important constituent; cosmological decades 40 through 100.

Brown dwarf. A stellar object that has too little mass to produce sustained nuclear fusion in its central core and is supported in part by degeneracy pressure.

Cambrian explosion. An intense burst of speciation that occurred on Earth 540 million years ago.

Chandrasekhar mass. The maximum mass of a white dwarf (or neutron star) that can be supported against gravitational collapse by degeneracy pressure.

Closed universe. A universe which contains enough energy density to halt its expansion and recollapse into a "big crunch."

Cold dark matter. Any type of candidate for the dark matter particles with relatively large masses (usually greater than the proton mass) so that the particles are slowly moving. Weakly interacting massive particles are one leading candidate for cold dark matter.

Copernican revolution. The idea that Earth does not occupy a special location in space.

Copernican time principle. The idea that the current cosmological epoch has no special place in time.

Cosmic background radiation. The diffuse sea of radiation left over from the big bang.

Cosmological constant problem. The problem of why the cosmological constant, the vacuum energy density of the universe, has such a small value (its value is 120 orders of magnitude smaller than that suggested by the simplest arguments from particle physics).

Cosmological decade. A logarithmic unit of time used to measure very long time scales. If a time τ in years is written in scientific notation $\tau = 10^{\eta}$ years, then the exponent η is the number of cosmological decades.

Cosmological heat death. A type of effective heat death in which the expanding universe becomes exactly adiabatic so that no more entropy can be created. In such a universe, with no entropy generation, no interesting physical processes can take place and the universe becomes dull and lifeless.

Cosmological principle. The statement that the universe is both homogeneous and isotropic.

Cosmology. The study of the origin and evolution of the universe as a whole.

Dark Era. Time period in the future of the universe after stars are gone and black holes have evaporated; cosmological decades greater than 100. Only the waste products remain: electrons, positrons, neutrinos, and radiation.

Dark matter. Matter in the universe that emits no light (or very little light). A large fraction of the mass of the universe is in this form, which is detected only indirectly through its gravitational effects. For example, the halos of galaxies contain a large amount of dark matter.

Degeneracy pressure. The pressure produced in a very dense gas due to the quantum mechanical uncertainty principle. The wavelike nature of particles prevents them from being squeezed too close together and thereby results in a pressure. In this state, the pressure depends only on the gas density and not the temperature.

Degenerate Era. Time period in the future of the universe when the most important constituents are degenerate stellar objects left over from stellar evolution: brown dwarfs, white dwarfs, neutron stars, and black holes; cosmological decades 15 through 39.

Dynamical relaxation. Process by which a self-gravitating system, such as a star cluster or a galaxy, changes its structure. Gravitational scattering encounters change the orbits of the constituent stars (or other bodies), and these changes accumulate to alter the overall structure of the system.

Ediacara fauna. Collection of soft-bodied creatures which first appeared on Earth about 800 million years ago.

Electromagnetic force. One of the four forces of nature. This force includes both the force between charged particles and that due to magnetic fields.

Electromagnetic radiation. This radiation includes ordinary visible light waves, as well as that of other wavelengths: gamma rays, X rays, ultraviolet, infrared, and radio waves. In all cases, the radiation is a self-propagating disturbance involving oscillating electric and magnetic fields.

Embedding diagram. A visualization technique used to show the curvature of space in general relativity. In such diagrams, space is considered to be a two-dimensional surface, with the surface curvature representing the space-time curvature.

Entropy. In thermodynamics, a fundamental quantity that provides a measure of the amount of disorder in a physical system. The second law of thermodynamics states that the *total* amount of entropy in any isolated physical system either increases or stays the same.

Escape speed. The speed required to overcome the gravitational pull of an astronomical body and leave its surface.

Eternal inflation. In this version of the inflationary universe paradigm, at all times, some region of space-time, somewhere in the multiverse, is experiencing an inflationary phase of expansion.

Eukaryotic cells. Cells that are characterized by nucleus, mitochondrion, and/or chloroplasts, and are usually capable of mitotic cell division. Organisms with this cell type include most of the familiar large life forms on Earth, including plants, animals, and fungi.

Event horizon. The boundary that separates a black hole from the rest of the universe. This imaginary surface is the same as the Schwarzschild radius if the black hole is not spinning, and smaller than the Schwarzschild radius for a spinning black hole.

Final mass function. The distribution of masses of degenerate stellar remnants after stellar evolution has run its course. These objects include neutron stars, white dwarfs, and brown dwarfs.

Fission. Nuclear reaction in which a large nucleus is split apart into smaller nuclei, usually with extra neutrons as well.

Flat universe. A universe with the critical value of the energy density (more density than an open universe, but less than a closed universe). Such a universe expands forever, but at an ever decreasing rate of expansion.

Flatness problem. A problem facing a universe without inflation. In order to produce a universe as large and flat as our universe, the initial conditions must be very special; in particular, the density of the early universe must be equal to the critical value to a tremendous accuracy.

Fusion. Nuclear reaction in which two or more nuclei combine to form a larger (heavier) nucleus. Fusion reactions provide the energy source for ordinary stars and are often called nuclear burning.

General relativity. A comprehensive theory of space, time, and mass. As first developed by Albert Einstein, general relativity holds that gravitation is an effect of the curvature of the space-time continuum.

Globular cluster. Very dense star cluster; some of these globular clusters are among the oldest objects in the universe and are thought to contain the oldest stars.

Grand unification. At a very high energy scale, about 10^{16} GeV or 10^{29} degrees kelvin, the strong, weak, and electromagnetic forces become unified into one. These forces are three of the four fundamental forces of nature.

Gravitational radiation. Wavelike disturbances in the background space-time; according to the theory of general relativity, accelerating masses radiate energy in this manner.

Graviton. The massless particle that mediates the gravitational force. The graviton plays the analogous role in gravity that the photon plays in electromagnetism.

Greenhouse effect. A mechanism in which the introduction of certain gases into a planetary atmosphere leads to more heat being retained and hence a larger surface temperature on the planet.

Hadrons. The class of particles that are made up of quarks and antiquarks. The subclass of these hadronic particles that contain three quarks are the baryons, whereas the subclass of particles that contain a quark and an antiquark are called mesons.

Hawking radiation. The energy emitted by a black hole due to quantum mechanical effects. This process causes black holes to evaporate on very long time scales.

Heat death. The concept that once the universe is in complete thermodynamic equilibrium, no more work can be done. If heat death occurs, the universe would be a dull and lifeless place.

Helium flash. A tremendous burst of energy that occurs near the end of a star's lifetime. This energy is the result of thermonuclear fusion processes that convert helium into carbon over a relatively short period of time.

Hertzsprung-Russell diagram. A graph used to study stellar evolution. The stellar luminosity is plotted on the vertical axis and the surface temperature is plotted on the horizontal axis. Stars trace out well defined tracks in this diagram as they evolve.

Homogeneous. The same at every point in space. The universe is thought to be homogeneous on the largest scales.

Horizon problem. A problem facing a universe without inflation. All parts of our universe are observed to have nearly the same temperature, as determined by the cosmic background radiation, even though not all parts of the universe were in causal contact at earlier times (with no inflation).

Hot dark matter. A candidate for the dark matter particles with relatively small masses (usually about one-billionth of the proton mass) so that the particles move at relativistic speeds when their abundances are determined.

Hubble expansion. The overall expansion of the universe as predicted by big bang theory and measured by astronomers. From any given reference point, the expansion speed of distant galaxies increases with increasing distance; this relation is known as Hubble's law.

Inflationary universe. A modification of the big bang theory of the universe; in the early evolution of the universe, the expansion is rapidly accelerating.

Initial mass function. The distribution of stellar masses when stars are first formed.

Interstellar medium. The gas and dust that permeate the space between stars in a galaxy.

Isotropic. The same in all directions. The universe is thought to be isotropic on the largest scales.

Large-scale structure of the universe. The patterns formed by galaxies over immense distances.

Leptons. A particular class of elementary particles, including electrons, muons, tau particles, and their associated neutrinos. These particles are characterized by half integer spin, no color charge, and an approximately conserved property called lepton number.

Luminosity. The energy generation rate (power) of an astrophysical object.

Main sequence. In the Hertzsprung-Russell diagram, stars with the internal configurations appropriate for hydrogen fusion fall along a curve called the main sequence. Stars spend most of their lifetimes in this configuration.

Metastable state. A configuration of a physical system in which a high energy state is relatively long-lived. A lower energy state exists, but transitions to the lower energy state are inhibited by energy barriers, and hence the system retains its higher energy state for a long time.

Multiverse. A large region of space-time that includes an entire ensemble of different universes, each with its own properties.

Neutrino. An elementary particle with no charge and very little or no mass. Neutrinos interact only through the weak force (and gravity).

Neutron star. A small compact stellar object that is supported by degeneracy pressure of neutrons. These stellar remnants, with masses between one and two solar masses, are left over from the evolution and death of massive stars.

Nucleosynthesis. The creation of elements through nuclear reactions. Formation of the light elements takes place during the early history of the universe; formation of heavier elements takes place in stars.

Olbers' paradox. A problem facing a universe that is static, unchanging, and infinitely old. In such a universe, the night sky would appear as bright as a stellar surface.

Open universe. A universe that does not contain enough energy density to halt its expansion; an open universe continues to expand forever.

Panspermia. The notion that life on Earth originated elsewhere in the galaxy and was brought here by astrophysical bodies such as meteors, asteroids, or comets.

Phase transition. A change of state between two different phases or configurations of matter. Common examples include water freezing into ice, or water boiling into steam.

Photon. The particle that corresponds to electromagnetic radiation or light. Photons travel at the speed of light and have an energy that depends on their wavelength; the shorter the wavelength, the larger the energy.

Planck's constant. The fundamental constant of nature, often written as $\hbar = h/2\pi = 1.05 \times 10^{-27}$ erg·sec, that sets the scale for quantum mechanical processes.

Planck scale. Energy scale at which gravity becomes unified with the other three fundamental forces of nature. This energy is about 10^{19} GeV or 10^{32} degrees kelvin.

Positron. A positively charged antiparticle that is the partner to the electron; it is usually denoted as e^+.

Positronium. An atomlike structure consisting of an electron and a positron in orbit about each other.

Primordial black hole. A black hole produced in the early universe. Thought to be the smallest black holes, although these objects remain hypothetical.

Primordial Era. The earliest phase in the history of our universe, before the formation of astronomical structures of any kind. Before the universe was a million years old, it contained no stars, galaxies, or clusters.

Proton decay. Proposed process in which the proton decays into lighter particles such as photons, positrons, and neutrinos. The lifetime of a proton is very long, at least 10^{32} years, much longer than the current age of the universe.

Protostar. A star that is still in the process of forming by gaining mass from the interstellar medium.

Quantum gravity. Regime of physics in which both quantum mechanics and the theory of general relativity (strong gravity) are required to describe nature. Physicists are still working on a self-consistent theory of quantum gravity.

Quantum mechanics. Theory of physics that describes matter as having a wavelike character on very small length scales (usually the size of atoms and smaller).

Quarks. The fundamental constituent particles that make up protons, neutrons, and other composite particles known as hadrons.

Quasar. A young galaxy with an energetic luminosity source, powered by a supermassive black hole.

Radiation Dominated Era. An early phase in the history of the universe when the energy density of radiation was greater than that of ordinary matter. This phase lasted for the first few thousand years.

Recombination. The epoch in the history of the universe when electrons first combined with nuclei to form ordinary atoms; this event occurred when the universe was about 300,000 years old.

Red dwarf. Another name for a star of low mass, about 10 to 40 percent the mass of the Sun. Red dwarf stars are the most numerous and the longest-lived of all possible stars.

Redshift. If a source of light, such as a star or galaxy, moves away from us, its radiation is shifted toward lower frequencies or longer wavelengths (toward the red end of the spectrum).

Rest energy. The energy contained in the mass of an object when it is not moving. Also known as the rest mass. If the object can completely annihilate, the released energy is $E = mc^2$.

Schwarzschild radius. The effective outer boundary of a black hole. An object of a given mass must be compressed to this radial size to become a black hole.

Singularity. A point in space-time at which the density becomes infinite (as well as the temperature and pressure). The predictive power of physical law breaks down at such points. In the universe, singularities arise at the centers of black holes, at the very beginning of a big bang universe, and at the very last instant of a recollapsing universe.

Stellar black hole. A black hole with a mass comparable to that of a star, most likely in the range of three to thirty solar masses. Such black holes can be produced by a supernova explosion resulting from the death of a massive star.

Stellar remnants. Collectively refers to all of the end products of stellar evolution, including black holes, neutron stars, white dwarfs, and sometimes brown dwarfs as well.

Stelliferous Era. Time period in the evolution of the universe when stars provide the most important source of energy. We are currently living in this era, which spans cosmological decades 6 through 14.

Strong nuclear force. One of the four forces of nature. The strong force holds protons and neutrons together in atomic nuclei and plays an important role in nuclear fusion.

Supermassive black hole. Large black holes with millions to billions of solar masses of material. These gigantic black holes live in the centers of most galaxies. They provide the driving engine for active galactic nuclei and quasars.

Supernova. Violent explosion of an evolved star at the end of its nuclear burning life.

Tidal forces. For a given object, the tidal force is the difference between the gravitational force exerted on the near side of the object and that on the far side.

Time dilation. A relativistic effect in which time runs slower. Time dilation occurs for objects moving close to the speed of light and for objects near the surface of a black hole.

Uncertainty principle. Property of quantum systems (and other wavelike physical systems). States that the momentum and position of any particle cannot be known with absolute certainty.

Vacuum energy density. The energy associated with empty space. In principle, this energy can be substantial and hence "empty space" is not really so empty. This type of energy can create a repulsive gravitational force and drive the universe into a phase of inflationary expansion.

Virtual particles. Particles that arise from the quantum mechanical uncertainty principle. Such particles live for only a very short time.

Weak nuclear force. One of the four forces of nature. The weak force mediates radioactive decays and some nuclear fusion processes.

White dwarf. A stellar remnant supported by electron degeneracy pressure. Stars in the mass range between one-tenth and eight times the mass of the Sun end their lives as white dwarfs.

Weakly interacting massive particles. Particles that are thought to make up some fraction of the dark matter in the universe. These particles interact only through gravity and the weak force; their mass is expected to be approximately ten times the mass of the proton.

Wormhole. In general relativity theory, this proposed structure makes a connection between two black holes, either in separate universes or in separate regions of our universe. The wormhole effectively creates a bridge between the two black holes and hence a bridge between two different regions of space-time.

NOTES

Rather than encumber the flow of the story with footnotes and references, we provide a quick scientific review of the various issues for each chapter. In the collection of notes given below, we present a brief description of the relevant issues and the corresponding references. The citations refer to the reference list that follows. In some cases, for brevity, we list only representative references.

INTRODUCTION

Much of the material in this book is based on a review paper that outlines the physics of the future universe (Adams and Laughlin 1997, hereafter AL97; see also Adams and Laughlin 1998). An assortment of previous papers have dealt with the future of the universe. Rees (1969) considered the fate of a closed universe, whereas Islam (1977, 1979) and Dyson (1979) took up the case of an open or flat universe that expands forever. A series of other papers considers specific issues, such as the implications of proton decay (Feinberg 1981; Dicus et al. 1982; Turner 1983) and the formation of positronium (Page and McKee 1981ab). A general overview of the subject is provided by Islam (1983) and by Davies (1994).

A key question is whether the universe will expand forever, or at least live long enough to experience the time line presented in this book. Current astronomical data indicate that the density is less than (or perhaps equal to) the critical density, and hence continued expansion does lie in our future path (see the recent review by Dekel, Burstein, and White 1997, and references therein).

Many introductory textbooks explain the basics of astronomy and the four forces of nature (e.g., Shu 1982; see also Zuckerman and Malkan 1996). The conflict between gravity and thermodynamics is also outlined in Shu (1982).

The review paper of AL97 introduced the concepts of cosmological decades, the Copernican time principle, and the eras of the future universe. In standard big bang cosmology, the past history of the universe is usually broken up into the Radiation Dominated Era and the Matter Dominated Era. The transition between these eras occurs when the universe is about two thousand years old (where the exact number, like all cosmological parameters presented in this book, is subject to some uncertainty). These time periods are determined by the nature of the cosmological expansion. In this treatment, however, the time periods are determined by the inventory of astronomical objects. The Primordial Era, with no stellar objects or astronomical structures, is almost synonymous with the Radiation Dominated Era, but extends into the Matter Dominated Era until the first stars form.

The scaling hypothesis for life-forms was introduced in Dyson (1979).

The often-told story of Copernicus and Bruno can be found many places (see, e.g., the collection of Knickerbocker 1927). The formation of planets around other stars has been discussed for centuries (Kant 1755; Laplace 1796), but planets orbiting nearby stars were only recently discovered (Mayor and Queloz 1995; Marcy and Butler 1996; see also Marcy and Butler 1998 for a current review of the subject, and Croswell 1997 for a general treatment).

1. THE PRIMORDIAL ERA

This chapter discusses the current version of the big bang theory, including an inflationary epoch. For a comprehensive review of modern cosmology, see Kolb and Turner (1990). For a classic popular-level treatment of big bang cosmology, see Weinberg (1977). For a critical discussion of current cosmological issues, see the conference volume edited by Turok (1997).

The inflationary universe was introduced in Guth (1981). Other important early papers include: Albrecht and Steinhardt (1982); Linde (1982,

1983a); Bardeen, Steinhardt, and Turner (1983); Guth and Pi (1982); Steinhardt and Turner (1984). Comprehensive textbook treatments of inflation are given in both Linde (1990) and Kolb and Turner (1990). A popular-level treatment is provided in Guth (1997).

During inflation, when "points of space rush away from each other faster than the speed of light," we mean more precisely that the scale factor $R(t)$ grows faster than a linear function of time t (which includes the usual form $R(t) \propto e^{Ht}$).

The cosmic background radiation was discovered by Penzias and Wilson (1965). Two decades later, the COBE satellite showed that the spectrum was exceedingly close to a blackbody and then discovered small temperature fluctuations $\Delta T/T \sim 10^{-5}$ (Wright et al. 1992; Smoot et al. 1992). Follow-up observations from ground-based observatories have provided additional measurements of fluctuations of the cosmic background radiation on smaller angular scales (e.g., Meyer et al. 1991; Gaier et al. 1992; Shuster et al. 1993).

Baryogenesis is the process by which an excess of matter over antimatter is generated. The basic ingredients of baryon number violation, out of equilibrium reactions, and no time reversal were first outlined by Sakharov (1967). A more current review is given in Dolgov (1992).

Big bang nucleosynthesis began with Alpher, Bethe, and Gamow (1948; see also Gamow 1946), continued with Wagoner (1973), and then rapidly became a detailed enterprise (e.g., Walker et al. 1991). A good textbook treatment is provided by Kolb and Turner (1990).

The mass measurements in galactic halos and galaxy clusters, in conjunction with the results of big bang nucleosynthesis, make a compelling case for the existence of nonbaryonic dark matter (see, e.g., the review of Krauss 1986). Although the general properties of such matter are reasonably well constrained, the dark matter has not yet been identified (see, e.g., Diehl et al. 1995; Jungman et al. 1996; Spooner 1997).

2. THE STELLIFEROUS ERA

Although the formation of galaxies is an ongoing area of study, the basic principles are in place and can be found in most advanced textbooks (Peebles 1993; Kolb and Turner 1990). Similarly, the study of star formation is a rapidly advancing field. The current paradigm of the star formation process has been in place for over a decade (e.g., Shu, Adams, and Lizano 1987), and further progress continues (see the recent conference volume edited by Boss et al. 1999).

This chapter deals with many issues of stellar evolution, a science that has become well developed in the latter part of this century. Many of the topics discussed here are covered in graduate level textbooks (Clayton 1983; Kippenhahn and Weigert 1990; Hansen and Kawaler 1994; see also Chandrasekhar 1939).

The long-term fate of Earth depends crucially on the mass loss from the red giant Sun (see Sackmann et al. 1993). The long-term fate of the smallest stars, red dwarfs, has only recently been determined (Laughlin, Bodenheimer, and Adams 1997).

The calculations of a red dwarf entering our solar system and either scattering Earth into interstellar space, or capturing the planet, are previously unpublished. This result is related to scattering calculations of solar systems in stellar clusters (Laughlin and Adams 1998); this type of scattering may explain some of the orbits observed in extrasolar planetary systems (Marcy and Butler 1998).

A much more detailed account of the history of life on our planet is given in Schopf (1992). The oldest unambiguous fossils are 3.5 billion years old and are found in rock formations in Swaziland (South Africa) and Pilbara (Western Australia). Even older sedimentary rock formations are found in the Isua Supracrustal Group (Greenland), although these rocks have been severely metamorphosed and hence cannot show a clear-cut record of early life. The earliest known eukaryotes appeared in the fossil record about 1750 million years ago, diversified rapidly about 1100 million years ago, and reached a peak in abundance and diversity about 900 million years ago (for further detail, see Schopf 1992 and references therein).

In the discussions concerning the search for extraterrestrial life and the colonization of the galaxy, we are on rather speculative ground, much more so than in our discussions of physical phenomena. The quote from A. Comte (1835) was taken from the treatment of Pais (1986, p. 165).

The long-term prospects for star formation in the galaxy are estimated from studies of the star formation history in our galaxy and others (see, e.g., Kennicutt, Tamblyn, and Congdon 1994; Rana 1991; AL97). The upcoming metallicity increase of the galaxy is estimated in Timmes (1996).

3. THE DEGENERATE ERA

The inventory of the Degenerate Era is set by the combination of the initial mass function (IMF) for stars and the transformation between initial stellar masses and the masses of their degenerate remnants. The IMF remains a subject of current research, but is now understood in broad general terms (Salpeter 1955; Miller and Scalo 1979; Scalo 1986; Rana 1991; Adams and Fatuzzo 1996). The transformation between progenitor masses and remnant masses is relatively well known (see, e.g., Wood 1992), but the amount of mass loss experienced during the red giant phases requires further specification. The lowest-mass stellar objects—brown dwarfs—have only recently been discovered (compare Oppenheimer et al. 1995 and Golimowski et al. 1995 with older reviews of Stevenson 1991 and Tinney 1995), but are relatively well understood as astrophysical entities (Burrows et al. 1993; Burrows and Liebert 1993).

The dynamics of galaxy collisions are discussed in Binney and Tremaine (1987) and M. Weinberg (1989). Regarding our upcoming collision with Andromeda, the orbits of nearby galaxies are currently being measured (Peebles 1994; Riess et al. 1995). Dynamical relaxation of the galaxy is analogous to the dynamical relaxation of stellar clusters (see Binney and Tremaine 1987; Shu 1982; Lightman and Shapiro 1978); these latter systems are much smaller and change their structure on much shorter time scales so that these dynamical issues can be studied more directly.

Relatively little work has been done on direct stellar collisions because they are so rare in the present-day universe. The numerical simulation shown in Chapter 3 was taken directly from our review article (AL97). For discussions of helium and carbon burning stars, see Kippenhahn and Weigert (1990).

Although the exact nature of the nonbaryonic component of the dark matter has not been identified, its general properties are relatively well constrained (Diehl et al. 1995; Jungman et al. 1996; Spooner 1997). In particular, in order to have a cosmologically interesting abundance today, the dark matter interaction cross section must be of order $\sigma \sim 10^{-37}$ cm^2 (Kolb and Turner 1990) and hence white dwarfs will capture dark matter particles that stream through the stellar interior (AL97). The capture of dark matter in the Sun and Earth has also been studied (Faulkner and Gilliland 1985; Press and Spergel 1985; Krauss, Srednicki, and Wilczek 1986; Gould 1987).

Speculations about life in white dwarf atmospheres follow directly from the scaling hypothesis introduced by Dyson (1979); the speculations about life outside white dwarfs follow from simple accounting.

Although proton decay is predicted theoretically, experiments have thus far only set a lower limit on the proton lifetime of about 32 cosmological decades (Particle Data Group 1998; Langacker 1981; Perkins 1984). For the sake of definiteness, we adopt a proton lifetime of 37 cosmological decades for most of this discussion; other values for the proton lifetime can easily be accommodated without qualitatively changing the story. If the proton does not decay through the simplest channels predicted by grand unified theories (see, e.g., Langacker 1981; Kane 1993), a host of other proton decay channels are possible (e.g., Feinberg, Goldhaber, and Steigman 1978; Wilczek and Zee 1979; Mohapatra and Marshak 1980; Weinberg 1980; Goity and Sher 1995). In addition, the vacuum structure of electroweak theory allows for the nonconservation of baryon number; tunneling events between different vacuum states can give rise to a change in baryon number and the decay of protons with an expected time scale of about 140 cosmological decades (see Rajaraman 1987; Kolb and Turner 1990; 't Hooft 1976; AL97). Finally, gravitational effects can also drive proton decay with an expected lifetime from 45 to 169 cosmological decades (e.g., Zel'dovich 1976; Hawking, Page, and Pope 1979; Page 1980; Hawking 1987; see also Adams et al. 1998).

The effects of proton decay on stellar structure and evolution are discussed in Feinberg (1981), Dicus et al. (1982), Turner (1983), AL97, and Adams et al. (1998). Stellar remnants are also affected by other long-term processes such as pycnonuclear reactions (Shapiro and Teukolsky 1983; Salpeter and van Horn 1969) and spallation (Hubell, Grimm, and Overbo 1980).

4. THE BLACK HOLE ERA

The basic properties of black holes are outlined in many textbooks (Weinberg 1972; Misner, Thorne, and Wheeler 1973; Wald 1984; Ohanian and Ruffini 1994). Gravitational radiation is covered in these same texts. An especially good popular-level treatment of black holes and general relativity is given by Thorne (1994).

Observational evidence for black holes can be found in three different settings: the three-million-solar-mass black hole in the center of our galaxy (Genzel et al. 1996), supermassive black holes in the centers of external galaxies (Kormendy et al. 1997), and stellar mass black holes within our galaxy (Narayan et al. 1997). Thus far, no evidence for primordial black holes has been found (Carr 1976).

The dynamics of a black hole disrupting our solar system were calculated explicitly for this book, and have not appeared elsewhere.

The emission of radiation from black holes was first predicted over two decades ago (Hawking 1974, 1975). Although it remains to be discovered, and hence remains a purely theoretical concept, Hawking radiation is predicted in general terms and is covered in many textbooks (e.g., Wald 1984, 1994; Thorne et al. 1986; Birrell and Davies 1982). Because we lack a complete theory of quantum gravity, however, the final moments of a black hole's life remain controversial (e.g., Russo, Susskind, and Thorlacius 1992).

To the best of our knowledge, the theoretical construction of a black hole computer is original. However, the basic idea of constructing logic gates from unfamiliar materials has been used in other contexts (e.g., Poundstone 1985).

The question of whether the universe can make black holes faster than they evaporate remains open (see, e.g., Rees 1984, 1997; see also Page and McKee 1981a, Frautschi 1982).

5. THE DARK ERA

The inventory of the Dark Era follows directly from the contents of the previous cosmological eras (see also Page and McKee 1981ab; Barrow and Tipler 1986). The radiation backgrounds of the future universe are calculated in

AL97; the largest uncertainties arise from our lack of knowledge of the proton lifetime and the mass distribution of black holes.

Heat death of the universe has been widely discussed since the second law of thermodynamics became understood (Helmholtz 1854; Clausius 1865, 1868). In the context of big bang theory, the issue of heat death shifts to the question of adiabaticity (see, e.g., Tolman 1934; Eddington 1931; Barrow and Tipler 1978, 1986; Frautschi 1982; AL97). An important constraint on the long-term entropy production of the universe is given by the bound of Bekenstein (1981).

The formation and decay of positronium in the future universe is calculated in Page and McKee (1981ab). Continuing particle annihilation is covered in many sources (e.g., Frautschi 1982; Barrow and Tipler 1986; AL97).

Perhaps the most speculative physical process in this book is the possibility of a future phase transition, which can be driven by the quantum tunneling of a scalar field. The first such calculations were provided by Voloshin et al. (1975) and by Coleman (1977, 1985). These calculations have subsequently been generalized to include gravity (Coleman and De Luccia 1980), finite temperature effects (Linde 1983b), and more general forms of the scalar field potential (Adams 1993). Cosmological implications of vacuum phase transitions have also been discussed (Hut and Rees 1983; Turner and Wilczek 1982). If and when such a phase transition takes place, the laws of physics can change accordingly (Crone and Sher 1990; Sher 1989; Sukuki 1988; Primack and Sher 1980). A related process is the nucleation of child universes (Sato et al. 1982; Blau, Guendelman, and Guth 1987; Hawking 1987; Farhi, Guth, and Guven 1990; see also Guth 1997). Such child universes can in principle receive information from our universe, and perhaps even matter (for varying points of view, see, e.g., Visser 1995; Linde 1988, 1989; Tipler 1992).

CONCLUSION

Long-term expansion depends on a number of factors, including the vacuum contribution to the energy density (Weinberg 1989; Carroll, Press, and Turner 1992), the mass density of the universe (see Turok 1997 for a recent overview;

see also Loh and Spillar 1986), and various other considerations (e.g., Ellis and Rothman 1993; Gott 1993; Grischuk and Zel'dovich 1978).

Although current astronomical data suggest that the universe will continue to expand (Dekel et al. 1997), we briefly discuss the scenario in which the universe, or a portion of it, recollapses. For a classic treatment of the physical events leading up to the big crunch, see the paper of Rees (1969).

Fine-tuning of our universe is discussed in many contexts, and is related to the "anthropic cosmological principle" (for further discussion, see Barrow and Tipler 1986; Carr and Rees 1979). The concept of our universe being but one of many possible universes is now gaining considerable cosmological attention (e.g., Rees 1981); a recent general-level treatment of the multiverse and its implications is presented in Rees (1997).

The idea of eternal inflation and eternal complexity has been widely discussed by A. Linde (see, e.g., Linde 1986, 1988, 1989, 1990, 1994; see also Vilenkin 1983). Darwinian evolution of universes has been introduced by L. Smolin and is described in his recent book (Smolin 1997; see also Rees 1997).

REFERENCES AND FURTHER READING

Adams, F. C., 1993, General solutions for tunneling of scalar fields with quartic potentials, *Phys. Rev.* D **48,** 2800.

Adams, F. C., and M. Fatuzzo, 1996, A theory of the initial mass function for star formation in molecular clouds, *Astrophys. J.* **464,** 256.

Adams, F. C., and G. Laughlin, 1997, A dying universe: The long-term fate and evolution of astrophysical objects, *Rev. Mod. Phys.* **69,** 337.

Adams, F. C., and G. Laughlin, 1998, The future of the universe, *Sky and Telescope* **96,** 32.

Adams, F. C., G. Laughlin, M. Mbonye, and M. J. Perry, 1998, The gravitational demise of cold degenerate stars, *Phys. Rev.* D **58,** 083003.

Albrecht, A., and P. J. Steinhardt, 1982, Cosmology for grand unified theories with radiatively induced symmetry breaking, *Phys. Rev. Lett.* **48,** 1220.

Alpher, R. A., H. Bethe, and G. Gamow, 1948, The origin of chemical elements, *Phys. Rev.* **73,** 803.

Bahcall, J. N., 1989, *Neutrino Astrophysics* (Cambridge: Cambridge Univ. Press).

Bardeen, J. M., P. J. Steinhardt, and M. S. Turner, 1983, Spontaneous creation of almost scale-free density perturbations in an inflationary universe, *Phys. Rev.* D **28,** 679.

Barrow, J. D., and F. J. Tipler, 1978, Eternity is unstable, *Nature* **276,** 453.

Barrow, J. D., and F. J. Tipler, 1986, *The Anthropic Cosmological Principle* (Oxford: Oxford Univ. Press).

Bekenstein, J. D. 1981, A universal upper bound to the entropy to energy ratio for bounded systems, *Phys. Rev.* D **23**, 287.

Binney, J., and S. Tremaine, 1987, *Galactic Dynamics* (Princeton: Princeton Univ. Press).

Birrell, N. D., and P. C. W. Davies, 1982, *Quantum Fields in Curved Space* (Cambridge: Cambridge Univ. Press).

Blau, S. K., E. I. Guendelman, and A. H. Guth, 1987, Dynamics of false-vacuum bubbles, *Phys. Rev.* D **35**, 1747.

Bond, J. R., B. J. Carr, and C. J. Hogan, 1991, Cosmic backgrounds from primeval dust, *Astrophys. J.* **367**, 420.

Boss, A., V. Mannings, and S. Russell, 1999, editors, *Protostars and Planets IV* (Tucson: Univ. Arizona Press).

Burrows, A., W. B. Hubbard, D. Saumon, and J. I. Lunine, 1993, An expanded set of brown dwarf and very low mass star models, *Astrophys. J.* **406**, 158.

Burrows, A., and J. Liebert, 1993, The science of brown dwarfs, *Rev. Mod. Phys.* **65**, 301.

Carr, B. J., 1994, Baryonic dark matter, *Ann. Rev. Astron. Astrophys.* **32**, 531.

Carr, B. J., 1976, Some cosmological consequences of primordial black hole evaporation, *Astrophys. J.* **206**, 8.

Carr, B. J., and M. J. Rees, 1979, The anthropic principle and the structure of the physical world, *Nature* **278**, 605.

Carroll, S. M., W. H. Press, and E. L. Turner, 1992, The cosmological constant, *Ann. Rev. Astron. Astrophys.* **30**, 499.

Chandrasekhar, S., 1939, *Stellar Structure* (New York: Dover).

Clausius, R., 1865, *Ann. Physik* **125**, 353.

Clausius, R., 1868, *Phil. Mag.* **35**, 405.

Clayton, D. D., 1983, *Principles of Stellar Evolution and Nucleosynthesis* (Chicago: Univ. Chicago Press).

Coleman, S., 1977, The fate of the false vacuum: 1. Semiclassical theory, *Phys. Rev.* D **15**, 2929.

Coleman, S., 1985, *Aspects of Symmetry* (Cambridge: Cambridge Univ. Press).

Coleman, S., and F. De Luccia, 1980, Gravitational effects on and of vacuum decay, *Phys. Rev.* D **21**, 3305.

Comte, A., 1835, *Cours de la Philosophie Positive*, **2**, 2 (Paris: Bachelier; repr. by Editions Anthropos, Paris 1968).

Crone, M. M., and M. Sher, 1990, The environmental impact of vacuum decay, *Am. J. Phys.* **59**, 25.

Croswell, K., 1997, *Planet Quest* (New York: The Free Press).

Davies, P.C.W., 1982, *The Accidental Universe* (Cambridge: Cambridge Univ. Press).

Davies, P.C.W., 1994, *The Last Three Minutes* (New York: Basic Books).

Dekel, A., D. Burstein, and S. D. M. White, 1997, Measuring omega, in *Critical Dialogues in Cosmology*, ed. N. Turok (Singapore: World Scientific), p. 175.

Dicus, D. A., J. R. Letaw, D. C. Teplitz, and V. L. Teplitz, 1982, Effects of proton decay on the cosmological future, *Astrophys. J.* **252**, 1.

Diehl, E., G. L. Kane, C. Kolda, and J. D. Wells, 1995, Theory, phenomenology, and prospects for detection of supersymmetric dark matter, *Phys. Rev.* D **52**, 4223.

Dolgov, A. D., 1992, Non-GUT baryogenesis, *Physics Reports* **222**, 309.

Dyson, F. J., 1979, Time without end: Physics and biology in an open universe, *Rev. Mod. Phys.* **51**, 447.

Dyson, F. J., 1988, *Infinite in All Directions* (New York: Harper and Row).

Eddington, A. S., 1931, *Nature* **127**, 447.

Ellis, G. F. R., and T. Rothman, 1993, Lost horizons, *Am. J. Phys.* **61**, 883.

Ellis, G. F. R., and D. H. Coule, 1994, Life at the end of the universe, *Gen. Rel. and Grav.* **26**, 731.

Farhi, E. H., A. H. Guth, and J. Guven, 1990, Is it possible to create a universe in the laboratory by quantum tunneling? *Nuclear Phys.* **B339**, 417.

Faulkner, J., and R. L. Gilliland, 1985, Weakly interacting massive particles and the Solar neutrino flux, *Astrophys. J.* **299**, 994.

Feinberg, G., 1981, The coldest neutron star, *Phys. Rev.* D **23**, 3075.

Feinberg, G., M. Goldhaber, and G. Steigman, 1978, Multiplicative baryon-number conservation and the oscillation of hydrogen into antihydrogen, *Phys. Rev.* D **18**, 1602.

Frautschi, S., 1982, Entropy in an expanding universe, *Science* **217**, 593.

Gaier, T., et al., 1992, A degree-scale measurement of anisotropy of the cosmic background radiation, *Astrophys. J. Lett.* **398**, L1.

Gamow, G., 1946, Expanding universe and the origin of elements, *Phys. Rev.* **70**, 572.

Genzel, R., et al., 1996, The dark mass concentration in the central parsec of the Milky Way, *Astrophys. J.* **472**, 153.

Goity, J. L., and M. Sher, 1995, Bounds on $\Delta B = 1$ couplings in the supersymmetric standard model, *Phys. Lett.* **346 B**, 69.

Golimowski, D. A., T. Nakajima, S. R. Kulkarni, and B. R. Oppenheimer, 1995, Detection of a very low mass companion to the astrometric binary Gliese 105A, *Astrophys. J. Lett.* **444**, L101.

Gott, J. R., 1993, Implications of the Copernican Principle for our future prospects, *Nature* **363**, 315.

Gould, A., 1987, Resonant enhancements in weakly interacting massive particle capture by the Earth, *Astrophys. J.* **321**, 571.

Grischuk, L. P., and Ya. B. Zel'dovich, 1978, Long wavelength perturbations of a Friedmann universe, and anisotropy of the microwave background, *Sov. Astron.* **22**, 125.

Guth, A., 1981, The inflationary universe: A possible solution to the horizon and flatness problems, *Phys. Rev.* D **23**, 347.

Guth, A., 1997, *The Inflationary Universe: The Quest for a New Theory of Cosmic Origins* (Reading, MA: Addison-Wesley).

Guth, A. H., and S.-Y. Pi, 1982, Fluctuations in the new inflationary universe, *Phys. Rev. Lett.* **49**, 1110.

Hansen, C. J., and S. D. Kawaler, 1994, *Stellar Interiors: Physical Principles, Structure, and Evolution* (New York: Springer).

Hawking, S. W., 1974, Black hole explosions? *Nature* **248**, 30.

Hawking, S. W., 1975, Particle creation by black holes, *Comm. Math. Phys.* **43**, 199.

Hawking, S. W., 1976, Black holes and thermodynamics, *Phys. Rev.* D **13**, 191.

Hawking, S. W., 1982, The development of irregularities in a single bubble inflationary universe, *Phys. Lett.* **115 B**, 295.

Hawking, S. W., 1987, Quantum coherence down the wormhole, *Phys. Lett.* **195 B**, 337.

Hawking, S. W., D. N. Page, and C. N. Pope, 1979, The propagation of particles in spacetime foam, *Phys. Lett.* **86 B**, 175.

Helmholz, H. von, 1854, *On the Interaction of Natural Forces.*

't Hooft, G., 1976, Symmetry breaking through Bell-Jackiw anomalies, *Phys. Rev. Lett.* **37**, 8.

Hubbell, J. H., H. A. Grimm, and I. Overbo, 1980, Pair, triplet, and total atomic cross sections for 1 MeV–100 GeV photons in elements $Z = 1$ to 100, *J. Phys. Chem. Ref. Data* **9**, 1023.

Hut, P., and M. J. Rees, 1983, How stable is our vacuum? *Nature* **302**, 508.

Islam, J. N., 1977, Possible ultimate fate of the universe, *Quart. J. R. Astron. Soc.* **18**, 3.

Islam, J. N., 1979, The ultimate fate of the universe, *Sky and Telescope* **57**, 13.

Islam, J. N., 1983, *The Ultimate Fate of the Universe* (Cambridge: Cambridge Univ. Press).

Jungman, G., M. Kamionkowski, and K. Griest, 1996, Supersymmetric dark matter, *Physics Reports* **267**, 195.

Kane, G. L., 1993, *Modern Elementary Particle Physics* (Reading, MA: Addison-Wesley).

Kane, G. L., 1995, *The Particle Garden* (Reading, MA: Addison-Wesley).

Kant, I., 1755, *Allegmeine Naturgeschichte und Theorie des Himmels.*

Kennicutt, R. C., P. Tamblyn, and C. W. Congdon, 1994, Past and future star formation in disk galaxies, *Astrophys. J.* **435**, 22.

Kippenhahn, R., and A. Weigert, 1990, *Stellar Structure and Evolution* (Berlin: Springer).

Knickerbocker, W. S., 1927, *Classics of Modern Science* (Boston: Beacon Press).

Kolb, E. W., and M. S. Turner, 1990, *The Early Universe* (Redwood City, CA: Addison-Wesley).

Kormendy, J., et al., 1997, Spectroscopic evidence for a supermassive black hole in NCG 4486B, *Astrophys. J.* **482**, L139.

Krauss, L., 1986, Dark matter in the universe, *Scientific American* **255**, 58.

Krauss, L. M., M. Srednicki, and F. Wilczek, 1986, Solar system constraints and signature for dark matter candidates, *Phys. Rev.* D **33**, 2079.

Langacker, P., 1981, Grand unified theories and proton decay, *Physics Reports* **72**, 186.

Laplace, P. S., 1796, *Exposition du systeme du monde* (Paris).

Laughlin, G., and F. C. Adams, 1998, The modification of planetary orbits in dense stellar clusters, *Astrophys. J. Lett.* **508**, L171.

Laughlin, G., P. Bodenheimer, and F. C. Adams, 1997, The end of the main sequence, *Astrophys. J.* **482**, 420.

Lightman, A. P., and S. L. Shapiro, 1978, The dynamical evolution of globular clusters, *Rev. Mod. Phys.* **50**, 437.

Linde, A. D., 1982, A new inflationary universe scenario: A possible solution of the horizon, flatness, homogeneity, isotropy, and primordial monopole problems, *Phys. Lett.* **108 B,** 389.

Linde, A. D., 1983a, Chaotic inflation, *Phys. Lett.* **129 B,** 177.

Linde, A. D., 1983b, Decay of the false vacuum at finite temperature, *Nucl. Phys.* **B216,** 421.

Linde, A. D., 1986, Eternally existing self-reproducing chaotic inflationary universe, *Phys. Lett.* **175B,** 395.

Linde, A. D., 1988, Life after inflation, *Phys. Lett.* **211 B,** 29.

Linde, A. D., 1989, Life after inflation and the cosmological constant problem, *Phys. Lett.* **227 B,** 352.

Linde, A. D., 1990, *Particle Physics and Inflationary Cosmology* (New York: Harwood Academic).

Linde, A., 1994, The self-reproducing inflationary universe, *Scientific American* **271,** 48.

Loh, E., and E. Spillar, 1986, A measurement of the mass density of the universe, *Astrophys. J. Lett.* **307,** L1.

Manchester, R. N., and J. H. Taylor, 1977, *Pulsars* (San Francisco: W. H. Freeman).

Marcy, G. W., and R. P. Butler, 1996, A planetary companion to 70 Virginis, *Astrophys. J. Lett.* **464,** L147.

Marcy, G. W., and R. P. Butler, 1998, Detection of extrasolar giant planets, *Ann. Rev. Astron. Astrophys.* **36,** 57.

Mayor, M., and D. Queloz, 1995, A Jupiter-mass companion to a solar-type star, *Nature* **378,** 355.

Meyer, S. S., E. S. Cheng, and L. A. Page, 1991, A measurement of the large-scale cosmic microwave background anisotropy at 1.8 millimeter wavelength, *Astrophys. J. Lett.* **410,** L57.

Mihalas, D., and J. Binney, 1981, *Galactic Astronomy: Structure and Kinematics* (New York: W. H. Freeman).

Miller, G. E., and J. M. Scalo, 1979, The initial mass function and stellar birthrate in the solar neighborhood, *Astrophys. J. Suppl.* **41,** 513.

Misner, C. W., K. S. Thorne, and J. A. Wheeler, 1973, *Gravitation* (San Francisco: W. H. Freeman).

Mohapatra, R. N., and R. E. Marshak, 1980, Local B-L symmetry of electroweak interactions, Majorana neutrinos, and neutron oscillations, *Phys. Rev. Lett.* **44,** 1316.

Narayan, R., D. Barret, and J. E. McClintock, 1997, Advection-dominated accretion model of black hole V404 Cygni in quiescence, *Astrophys. J.* **482**, 448.

Ohanian, H. C., and R. Ruffini, 1994, *Gravitation and Spacetime* (New York: W. W. Norton).

Oppenheimer, B. R., S. R. Kulkarni, K. Matthews, and T. Nakajima, 1995, The infrared spectrum of the cool brown dwarf G1229B, *Science* **270**, 1478.

Page, D. N., 1980, Particle transmutations in quantum gravity, *Phys. Lett.* **95 B**, 244.

Page, D. N., and M. R. McKee, 1981a, Matter annihilation in the late universe, *Phys. Rev.* D **24**, 1458.

Page, D. N., and M. R. McKee, 1981b, Eternity matters, *Nature* **291**, 44.

Pais, A., 1986, *Inward Bound* (Oxford: Oxford Univ. Press).

Particle Data Group, 1998, Particle physics booklet, *European Phys. J.* **C3**, 1.

Peebles, P. J. E., 1993, *Principles of Physical Cosmology* (Princeton: Princeton Univ. Press).

Peebles, P. J. E., 1994, Orbits of nearby galaxies, *Astrophys. J.* **429**, 43.

Penzias, A. A., and R. W. Wilson, 1965, A measurement of excess antenna temperature at 4080 Mc/s, *Astrophys. J.* **142**, 419.

Perkins, D., 1984, Proton decay experiments, *Ann. Rev. Nucl. Part. Sci.* **34**, 1.

Poundstone, W., 1985, *The Recursive Universe* (New York: Morrow).

Press, W. H., and D. N. Spergel, 1985, Capture by the Sun of a galactic population of weakly interacting massive particles, *Astrophys. J.* **296**, 679.

Primack, J. R., and M. Sher, 1980, Photon mass at low temperature, *Nature* **288**, 680.

Rajaraman, R., 1987, *Solitons and Instantons* (Amsterdam: North-Holland).

Rana, N. C., 1991, Chemical evolution of the galaxy, *Ann. Rev. Astron. Astrophys.* **29**, 129.

Rees, M. J., 1969, The collapse of the universe: An eschatological study, *Observatory* **89**, 193.

Rees, M. J., 1981, Our universe and others, *Quart. J. R. Astron. Soc.* **22**, 109.

Rees, M. J., 1984, Black hole models for active galactic nuclei, *Ann. Rev. Astron. Astrophys.* **22**, 471.

Rees, M. J., 1997, *Before the Beginning: Our Universe and Others* (Reading, MA: Addison-Wesley).

Riess, A. G., W. H. Press, and R. P. Kirshner, 1995, Determining the motion of the local group using type Ia supernova light curve shapes, *Astrophys. J. Lett.* **438**, L17.

Russo, J. G., L. Susskind, and L. Thorlacius, 1992, End point of Hawking radiation, *Phys. Rev.* D **46**, 3444.

Sackmann, I.-J., A. I. Boothroyd, and K. E. Kraemer, 1993, Our Sun III: Present and future, *Astrophys. J.* **418**, 457.

Sakharov, A. D., 1967, Violation of CP invariance, C asymmetry, and baryon asymmetry of the universe, *JETP Letters* **5**, 24.

Salpeter, E. E., 1955, The luminosity function and stellar evolution, *Astrophys. J.* **121**, 161.

Salpeter, E. E., and H. M. van Horn, 1969, Nuclear reaction rates in high densities, *Astrophys. J.* **155**, 183.

Sato, K., H. Kodama, M. Sasaki, and K. Maeda, 1982, Multiproduction of universes by first order phase transition of a vacuum, *Phys. Lett.* **108 B**, 103.

Scalo, J. M., 1986, The stellar initial mass function, *Fund. Cos. Phys.* **11**, 1.

Schopf, J., 1992, editor, *Major Events in the History of Life* (Boston: Jones and Bartlett).

Schuster, J., et al., 1993, Cosmic background radiation anisotropy at degree scales: Further results from the South Pole, *Astrophys. J. Lett.* **412**, L47.

Shapiro, S. L., and S. A. Teukolsky, 1983, *Black Holes, White Dwarfs, and Neutron Stars: The Physics of Compact Objects* (New York: Wiley).

Sher, M., 1989, Electroweak Higgs potentials and vacuum stability, *Physics Reports* **179**, 273.

Shu, F. H., 1982, *The Physical Universe* (Mill Valley, CA: University Science Books).

Shu, F. H., F. C. Adams, and S. Lizano, 1987, Star formation in molecular clouds: Observation and theory, *Ann. Rev. Astron. Astrophys.* **25**, 23.

Smolin, L., 1997, *Life of the Cosmos* (New York: Oxford Univ. Press).

Smoot, G., et al., 1992, Structure in the COBE differential microwave radiometer first-year maps, *Astrophys. J. Lett.* **396**, L1.

Spooner, N. J. C., 1997, editor, *The Identification of Dark Matter* (London: World Scientific).

Steinhardt, P. J., and M. S. Turner, 1984, A prescription for successful new inflation, *Phys. Rev.* D **29**, 2162.

Stevenson, D. J., 1991, The search for brown dwarfs, *Ann. Rev. Astron. Astrophys.* **29**, 163.

Suzuki, M., 1988, Slightly massive photon, *Phys. Rev.* D **38**, 1544.

Thorne, K. S., R. H. Price, and D. A. MacDonald, 1986, *Black Holes: The Membrane Paradigm* (New Haven: Yale Univ. Press).

Thorne, K. S., 1994, *Black Holes and Time Warps: Einstein's Outrageous Legacy* (New York: Norton).

Timmes, F. X., 1996, unpublished calculations.

Tinney, C. G., 1995, editor, *The Bottom of the Main Sequence and Beyond* (Berlin: Springer).

Tipler, F. J., 1992, The ultimate fate of life in universes which undergo inflation, *Phys. Lett.* **286 B,** 36.

Tolman, R. C., 1934, *Relativity, Thermodynamics, and Cosmology* (Oxford: Clarendon Press).

Turner, M. S., 1983, The end may be hastened by magnetic monopoles, *Nature* **306,** 161.

Turner, M. S., and F. Wilczek, 1982, Is our vacuum metastable? *Nature* **298,** 633.

Turok, N., 1997, editor, *Critical Dialogues in Cosmology* (Singapore: World Scientific).

Vilenkin, A., 1983, Birth of inflationary universes, *Phys. Rev.* D **27,** 2848.

Visser, M., 1995, *Lorentzian Wormholes: From Einstein to Hawking* (Woodbury, NY: AIP Press).

Voloshin, M. B., I. Yu. Kobzarev, and L. B. Okun, 1975, Bubbles in metastable vacuum, *Sov. J. Nucl. Phys.* **20,** 644.

Wagoner, R., 1973, Big bang nucleosynthesis revisited, *Astrophys. J.* **179,** 343.

Wald, R. M., 1984, *General Relativity* (Chicago: Univ. Chicago Press).

Wald, R. M., 1994, *Quantum Field Theory in Curved Spacetime and Black Hole Thermodynamics* (Chicago: Univ. Chicago Press).

Walker, T. P., G. Steigman, D. N. Schramm, K. A. Olive, and H.-S. Kang, 1991, Primordial nucleosynthesis redux, *Astrophys. J.* **376,** 51.

Weinberg, M. D., 1989, Self-gravitating response of a spherical galaxy to sinking satellites, *Mon. Not. R. Astron. Soc.* **239,** 549.

Weinberg, S., 1972, *Gravitation and Cosmology* (New York: Wiley).

Weinberg, S., 1977, *The First Three Minutes* (New York: Basic).

Weinberg, S., 1978, A new light boson? *Phys. Rev. Lett.* **40,** 223.

Weinberg, S., 1980, Varieties of baryon and lepton nonconservation, *Phys. Rev.* D **22,** 1694.

Weinberg, S., 1989, The cosmological constant problem, *Rev. Mod. Phys.* **61,** 1.

Weinberg, S., 1995, *Quantum Theory of Fields* (Cambridge: Cambridge Univ. Press).

Wilczek, F., and A. Zee, 1979, Conservation or violation of B-L in proton decay, *Phys. Lett.* **88 B**, 311.

Wood, M. A., 1992, Constraints on the age and evolution of the galaxy from the white dwarf luminosity function, *Astrophys. J.* **386**, 539.

Wright, E. L., et al., 1992, Interpretation of the cosmic microwave background radiation anisotropy detected by the COBE differential microwave radiometer, *Astrophys. J. Lett.* **396**, L13.

Zel'dovich, Ya. B., 1976, A new type of radioactive decay: Gravitational annihilation of baryons, *Phys. Lett.* **59 A**, 254.

Zuckerman, B., and M. A. Malkan, 1996, editors, *The Origin and Evolution of the Universe* (Sudbury, MA: Jones and Bartlett).

KEY EVENTS IN THE BIOGRAPHY OF THE UNIVERSE

EVENT	DECADE*
THE PRIMORDIAL ERA	
The big bang	$-\infty$
Quantum gravity rules universe (Planck epoch)	-50.5
Grand unification of three fundamental forces	-44.5
Quarks become confined into hadrons	-12.5
The first production of elements (nucleosynthesis)	-6
Matter energy dominates over radiation	4
Electrons and protons form atoms (recombination)	5.5
THE STELLIFEROUS ERA	
The first stars	6
Formation of the Milky Way	9
Formation of our solar system	9.5
Today	10
Our Sun dies	10.2
Close encounter of our galaxy with Andromeda	10.2
Death of the smallest stars	13
End of conventional star formation	14

(continued)

*The decades referred to here and throughout *The Five Ages of the Universe* are *cosmological* decades. When an interval of time is expressed in scientific notation, say 10^{η} years, then the exponent η is the cosmological decade.

(continued)

EVENT	DECADE
THE DEGENERATE ERA	
Planets become detached from stars	15
Brown dwarfs collide and form stars	16
Stars evaporate from our galaxy	19
Dark matter particles annihilate in the Galactic Halo	22.5
Gravitational radiation dissipates stellar orbits	24
White dwarfs deplete dark matter in the galactic halo	25
Black holes accrete Stars	30
Protons decay	37
Neutron stars lose mass and violently transform	38
Planets and white dwarfs are destroyed by proton decay	38–39
THE BLACK HOLE ERA	
Axions decay into photons	42
Hydrogen molecules experience cold fusion reactions	60
Stellar black holes evaporate	65–67
Black holes with one million solar masses evaporate	83
Positronium forms in a flat universe	85
The largest supermassive black holes disappear	98
THE DARK ERA	
Black hole with mass of current horizon scale evaporates	131
Positronium decays in a flat universe	141
Higher order proton decay processes occur	100–200
Cosmological phase transition reconstructs the universe	10–1000

INDEX

9 780684 865768